UGLY'S ELECTRICAL-REFERENCES

REVISED 1993

PRINTED IN U.S.A.

GEORGE V. HART

AND

SAMMIE HART

ALLIED PRINTING TRADES COUNCIL HOUSTON, TEXAS 31

united printing arts • 3509 Oak Forest Drive • Houston, Texas 77018 • (713) 688-6115

TABLE OF CONTENTS

TITLE		PAGE
OHM'S LAW		1 — 2
SERIES CIRCUITS		3 — 4
PARALLEL CIRCUITS		5 — 7
COMBINATION CIRCUITS		8 — 12
ELECTRICAL FORMULAS		13
TO FIND:	AMPERES (I)	14 — 19
	HORSEPOWER (HP)	20 — 21
	WATTS (P)	22
	KILO-WATTS (KW)	23 — 24
	KILO-VOLT-AMPERES (KVA)	25
	CAPACITANCE (C), AND CAPACITORS	26 — 29
	INDUCTION (L)	30
	IMPEDANCE (Z)	31
	REACTANCE (INDUCTIVE-XL, AND CAPACITIVE-XC)	32
	RESISTOR COLOR CODE	33
U.S. WEIGHTS AND MEASURES		34 — 35
METRIC SYSTEM		36 — 37
CONVERSION TABLES		38 — 39
METALS AND SPECIFIC RESISTANCE (K)		40 — 41
CENTIGRADE AND FAHRENHEIT THERMOMETER SCALES		42
USEFUL MATH, FORMULAS		43
THE CIRCLE		44
FRACTIONS		45 — 47
EQUATIONS		48 — 50
SQUARE ROOT		51
TRIGONOMETRY		52 — 53
CONDUIT BENDING		54 — 59
TAP, DRILL BIT, AND HOLE SAW TABLES		60
MOTORS:	RUNNING OVERLOAD UNITS	61
	BRANCH CIRCUIT PROTECTIVE DEVICES	62
	DIRECT CURRENT	63 — 65
	SINGLE-PHASE (A.C.)	
	TWO-PHASE (A.C.)	70 — 72
THREE-PHASE A.C. MOTORS		73 — 80
TRANSFORMERS:	CALCULATIONS	81
	VOLTAGE DROP CALCULATIONS	82
	SINGLE-PHASE CONNECTIONS	83
	BUCK AND BOOST CONNECTIONS	84
	FULL LOAD CURRENTS	85
	THREE-PHASE CONNECTIONS	86 — 90
	TWO-PHASE CONNECTIONS	91
	TWO-PHASE AND THREE-PHASE CONNECTIONS	92

TABLE OF CONTENTS (Continued)

TITLE	PAGE
MISCELLANEOUS WIRING DIAGRAMS	93 — 94
PROPERTIES OF CONDUCTORS	95
ALLOWABLE AMPACITIES OF CONDUCTORS	96 — 99
INSULATION CHARTS	100 — 103
MAXIMUM NUMBER OF CONDUCTORS IN CONDUIT	104 — 106
MAXIMUM NUMBER OF FIXTURE WIRES IN CONDUIT	107
TABLES: METAL BOXES	108
COVER REQUIREMENTS TO 600 VOLTS	109
VOLUME REQUIRED PER CONDUCTOR	109
CLEAR WORKING SPACE IN FRONT OF ELECTRICAL EQUIPMENT	110
MINIMUM CLEARANCE OF LIVE PARTS	111
GROUNDING	112 — 113
ELECTRICAL SYMBOLS	114 — 117
HAND SIGNALS FOR CRANES AND CHERRY PICKERS	118 — 119
USEFUL KNOTS	120
AMERICAN RED CROSS FIRST AID	121 — 124

OHM'S LAW

THE RATE OF THE FLOW OF THE CURRENT IS EQUAL TO ELECTROMOTIVE FORCE DIVIDED BY RESISTANCE.

ELECTROMOTIVE FORCE = VOLTS = "E"
CURRENT = AMPERES = "I"
RESISTANCE = OHMS = "R"

$$\text{AMPERES} = \frac{\text{VOLTS}}{\text{OHMS}}$$

SERIES CIRCUIT

A SERIES CIRCUIT IS A CIRCUIT THAT HAS ONLY ONE PATH THROUGH WHICH THE ELECTRONS MAY FLOW. NOTE: "T" STANDS FOR TOTAL.

ET = E1 + E2 + E3

IT = I1 = I2 = I3

RT = R1 + R2 + R3

PARALLEL CIRCUIT

A PARALLEL CIRCUIT IS A CIRCUIT THAT HAS MORE THAN ONE PATH THROUGH WHICH THE ELECTRONS MAY FLOW.

ET = E1 = E2 = E3

IT = I1 + I2 + I3

$$\frac{1}{RT} = \frac{1}{R1} + \frac{1}{R2} + \frac{1}{R3}$$

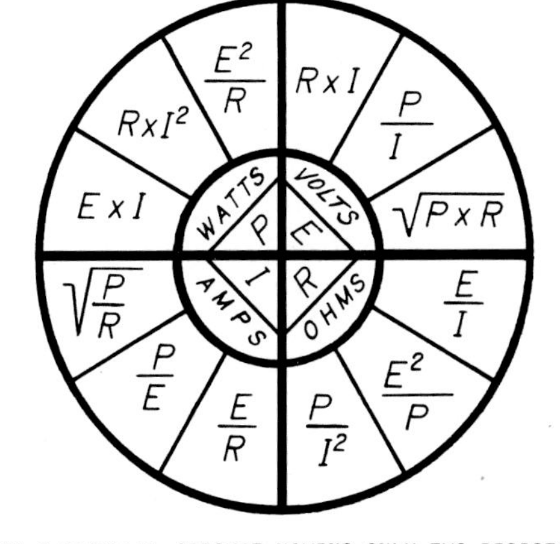

NOTE: FOR A PARALLEL CIRCUIT HAVING ONLY TWO RESISTORS, THE FOLLOWING FORMULA MAY BE USED.

$$RT = \frac{R1 \times R2}{R1 + R2}$$

OHM'S LAW

A. WHEN VOLTS AND OHMS ARE KNOWN:

$$\text{AMPERES} = \frac{\text{VOLTS}}{\text{OHMS}} \quad \text{OR} \quad I = \frac{E}{R}$$

EXAMPLE: FIND THE CURRENT OF A 120 VOLT CIRCUIT WITH A RESISTANCE OF 60 OHMS.

$$I = \frac{E}{R} = \frac{120}{60} = \underline{2 \text{ AMPERES}}$$

B. WHEN WATTS AND VOLTS ARE KNOWN:

$$\text{AMPERES} = \frac{\text{WATTS}}{\text{VOLTS}} \quad \text{OR} \quad I = \frac{P}{E}$$

EXAMPLE: A 120 VOLT CIRCUIT HAS A 1440 WATT LOAD, DETERMINE THE CURRENT.

$$I = \frac{P}{E} = \frac{1440}{120} = \underline{12 \text{ AMPERES}}$$

C. WHEN OHMS AND WATTS ARE KNOWN:

$$\text{AMPERES} = \sqrt{\frac{\text{WATTS}}{\text{OHMS}}} \quad \text{OR} \quad I = \sqrt{\frac{P}{R}}$$

EXAMPLE: A CIRCUIT CONSUMES 625 WATTS THROUGH A 12.75 OHM RESISTOR. DETERMINE THE CURRENT.

$$I = \sqrt{\frac{P}{R}} = \sqrt{\frac{625}{12.75}} = \sqrt{49} = \underline{7 \text{ AMPERES}}$$

NOTES:

A. ONE ELECTRICAL HORSEPOWER = 746 WATTS
ELECTRIC MOTORS ARE RATED IN HORSEPOWER.

B. ONE KILOWATT = 1000 WATTS
GENERATORS ARE RATED IN KILOWATTS.

SERIES CIRCUITS

RULE 1: THE TOTAL CURRENT IN A SERIES CIRCUIT IS EQUAL TO THE CURRENT IN ANY OTHER PART OF THE CIRCUIT.

TOTAL CURRENT = I(1) = I(2) = I(3), AND ETC.

RULE 2: THE TOTAL VOLTAGE IN A SERIES CIRCUIT IS EQUAL TO THE SUM OF THE VOLTAGES ACROSS ALL PARTS OF THE CIRCUIT.

TOTAL VOLTAGE = E(1) + E(2) + E(3), AND ETC.

RULE 3: THE TOTAL RESISTANCE OF A SERIES CIRCUIT IS EQUAL TO THE SUM OF THE RESISTANCES OF ALL THE PARTS OF THE CIRCUIT.

TOTAL RESISTANCE = R(1) + R(2) + R(3), AND ETC.

FORMULAS FROM OHM'S LAW

$$\text{AMPERES} = \frac{\text{VOLTS}}{\text{RESISTANCE}} \quad \text{OR} \quad I = \frac{E}{R}$$

$$\text{RESISTANCE} = \frac{\text{VOLTS}}{\text{AMPERES}} \quad \text{OR} \quad R = \frac{E}{I}$$

$$\text{VOLTS} = \text{AMPERES} \times \text{RESISTANCE} \quad \text{OR} \quad E = I \times R$$

EXAMPLE: FIND TOTAL VOLTAGE, TOTAL CURRENT, AND TOTAL RESISTANCE.

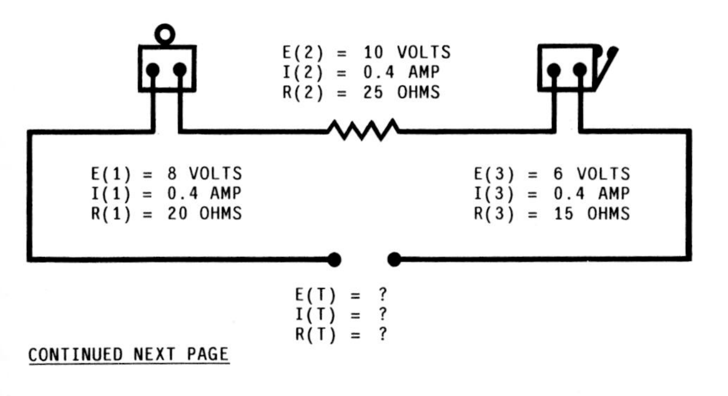

CONTINUED NEXT PAGE

SERIES CIRCUITS

$E(T) = E(1) + E(2) + E(3)$
$= 8 + 10 + 6$
$E(T) =$ 24 VOLTS

$I(T) = I(1) = I(2) = I(3)$
$= 0.4 = 0.4 = 0.4$
$I(T) =$ 0.4 AMP

$R(T) = R(1) + R(2) + R(3)$
$= 20 + 25 + 15$
$R(T) =$ 60 OHMS

EXAMPLE: FIND E(T), E(1), E(3), I(T), I(1), I(2), I(4), R(T), R(2), AND R(4). REMEMBER THAT THE TOTAL CURRENT IN A SERIES CIRCUIT IS EQUAL TO THE CURRENT IN ANY OTHER PART OF THE CIRCUIT.

E(1) = ?
I(1) = ?
R(1) = 72 OHMS

E(3) = ?
I(3) = 0.5 AMP
R(3) = 48 OHMS

E(2) = 12 VOLTS
I(2) = ?
R(2) = ?

E(4) = 48 VOLTS
I(4) = ?
R(4) = ?

E(T) = ?
I(T) = ?
R(T) = ?

$I(T) = I(1) = I(2) = I(3) = I(4)$
$I(T) = I(1) = I(2) = 0.5 = I(4)$
$0.5 = 0.5 = 0.5 = 0.5 = 0.5$
$I(T) =$ 0.5 AMP
$I(1) =$ 0.5 AMP
$I(2) =$ 0.5 AMP
$I(4) =$ 0.5 AMP

$E(1) = I(1) \times R(1)$
$= 0.5 \times 72$
$E(1) =$ 36 VOLTS

$E(T) = E(1) + E(2) + E(3) + E(4)$
$= 36 + 12 + 24 + 48$
$E(T) =$ 120 VOLTS

$E(3) = I(3) \times R(3)$
$= 0.5 \times 48$
$E(3) =$ 24 VOLTS

$R(T) = R(1) + R(2) + R(3) + R(4)$
$= 72 + 24 + 48 + 96$
$R(T) =$ 240 OHMS

$$R(2) = \frac{E(2)}{I(2)} = \frac{12}{0.5}$$

$R(2) =$ 24 OHMS

$$R(4) = \frac{E(4)}{I(4)} = \frac{48}{0.5}$$

$R(4) =$ 96 OHMS

PARALLEL CIRCUITS

RULE 1: THE TOTAL CURRENT IN A PARALLEL CIRCUIT IS EQUAL TO THE SUM OF THE CURRENTS IN ALL THE BRANCHES OF THE CIRCUIT.

TOTAL CURRENT = I(1) + I(2) + I(3), AND ETC.

RULE 2: THE TOTAL VOLTAGE ACROSS ANY BRANCH IN PARALLEL IS EQUAL TO THE VOLTAGE ACROSS ANY OTHER BRANCH AND IS ALSO EQUAL TO THE TOTAL VOLTAGE.

TOTAL VOLTAGE = E(1) = E(2) = E(3), AND ETC.

RULE 3: THE TOTAL RESISTANCE IN A PARALLEL CIRCUIT IS FOUND BY APPLYING OHM'S LAW TO THE TOTAL VALUES OF THE CIRCUIT.

$$\text{TOTAL RESISTANCE} = \frac{\text{TOTAL VOLTAGE}}{\text{TOTAL AMPERES}} \quad \text{OR} \quad RT = \frac{ET}{IT}$$

EXAMPLE: FIND THE TOTAL CURRENT, TOTAL VOLTAGE, AND TOTAL RESISTANCE.

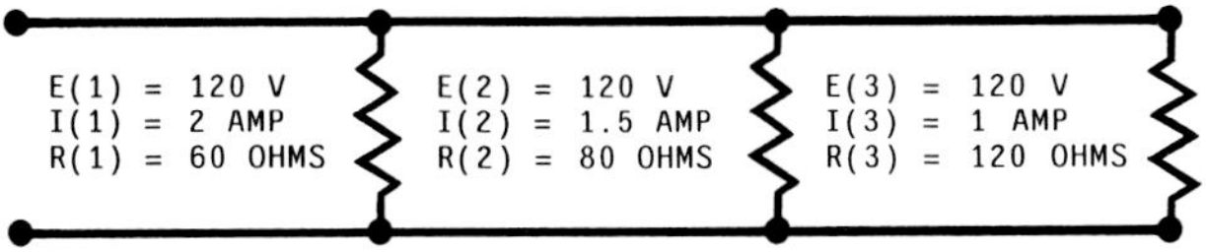

E(1) = 120 V
I(1) = 2 AMP
R(1) = 60 OHMS

E(2) = 120 V
I(2) = 1.5 AMP
R(2) = 80 OHMS

E(3) = 120 V
I(3) = 1 AMP
R(3) = 120 OHMS

I(T) = I(1) + I(2) + I(3)
= 2 + 1.5 + 1
I(T) = 4.5 AMP

E(T) = E(1) = E(2) = E(3)
= 120 = 120 = 120
E(T) = 120 VOLTS

$$R(T) = \frac{E(T)}{I(T)} = \frac{120 \text{ VOLTS}}{4.5 \text{ AMP}} = \text{26.66 OHMS RESISTANCE}$$

NOTE: IN A PARALLEL CIRCUIT THE TOTAL RESISTANCE IS ALWAYS LESS THAN THE RESISTANCE OF ANY BRANCH.

IF THE BRANCHES OF A PARALLEL CIRCUIT HAVE THE SAME RESISTANCE, THEN EACH WILL DRAW THE SAME CURRENT.

IF THE BRANCHES OF A PARALLEL CIRCUIT HAVE DIFFERENT RESISTANCES, THEN EACH WILL DRAW A DIFFERENT CURRENT.

IN EITHER SERIES OR PARALLEL CIRCUITS, THE LARGER THE RESISTANCE, THE SMALLER THE CURRENT DRAWN.

PARALLEL CIRCUITS

TO DETERMINE THE TOTAL RESISTANCE IN A PARALLEL CIRCUIT WHEN THE TOTAL CURRENT, AND TOTAL VOLTAGE ARE UNKNOWN.

$$\frac{1}{\text{TOTAL RESISTANCE}} = \frac{1}{R(1)} + \frac{1}{R(2)} + \frac{1}{R(3)} \quad \text{AND ETC.}$$

EXAMPLE: FIND THE TOTAL RESISTANCE.

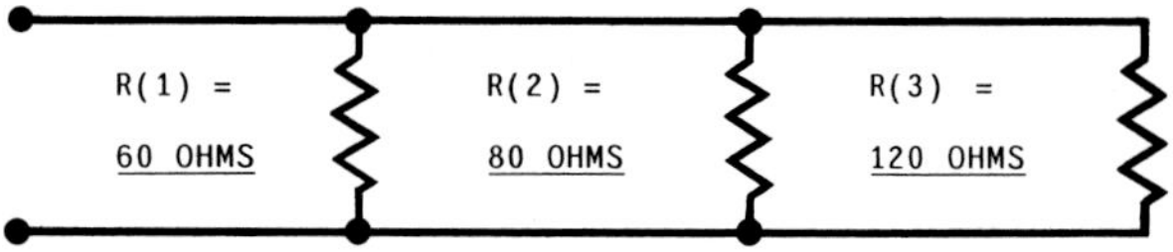

R(1) = 60 OHMS

R(2) = 80 OHMS

R(3) = 120 OHMS

$$\frac{1}{R(T)} = \frac{1}{R(1)} + \frac{1}{R(2)} + \frac{1}{R(3)}$$

$$\frac{1}{R(T)} = \frac{1}{60} + \frac{1}{80} + \frac{1}{120}$$

$$\frac{1}{R(T)} = \frac{4 + 3 + 2}{240}$$

USE LOWEST COMMON DENOMINATOR (240)

$$\frac{1}{R(T)} = \frac{9}{240}$$

CROSS MULTIPLY

$9 \times R(T) = 1 \times 240$ OR $9RT = 240$

DIVIDE BOTH SIDES OF THE EQUATION BY "9"

R(T) = 26.66 OHMS RESISTANCE

NOTE: THE TOTAL RESISTANCE OF A NUMBER OF EQUAL RESISTORS IN PARALLEL IS EQUAL TO THE RESISTANCE OF ONE RESISTOR DIVIDED BY THE NUMBER OF RESISTORS.

$$\text{TOTAL RESISTANCE} = \frac{\text{RESISTANCE OF ONE RESISTOR}}{\text{NUMBER OF RESISTORS IN CIRCUIT}}$$

CONTINUED NEXT PAGE

PARALLEL CIRCUITS

FORMULA: $R(T) = \frac{R}{N}$

EXAMPLE: FIND THE TOTAL RESISTANCE

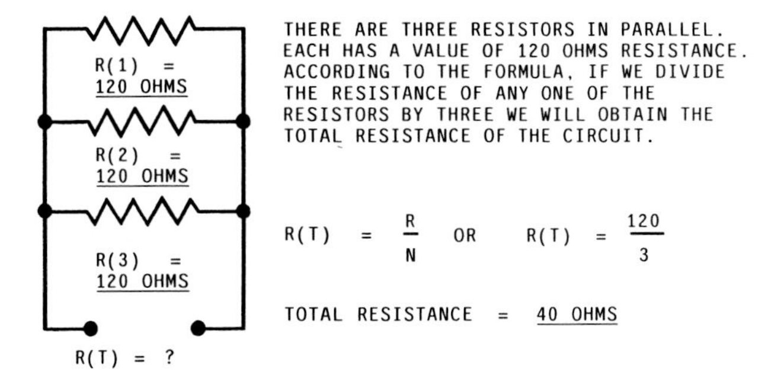

THERE ARE THREE RESISTORS IN PARALLEL. EACH HAS A VALUE OF 120 OHMS RESISTANCE. ACCORDING TO THE FORMULA, IF WE DIVIDE THE RESISTANCE OF ANY ONE OF THE RESISTORS BY THREE WE WILL OBTAIN THE TOTAL RESISTANCE OF THE CIRCUIT.

$R(T) = \frac{R}{N}$ OR $R(T) = \frac{120}{3}$

TOTAL RESISTANCE = 40 OHMS

NOTE: TO FIND THE TOTAL RESISTANCE OF ONLY TWO RESISTORS IN PARALLEL, MULTIPLY THE RESISTANCES, AND THEN DIVIDE THE PRODUCT BY THE SUM OF THE RESISTORS.

FORMULA: $\text{TOTAL RESISTANCE} = \frac{R(1) \times R(2)}{R(1) + R(2)}$

EXAMPLE:

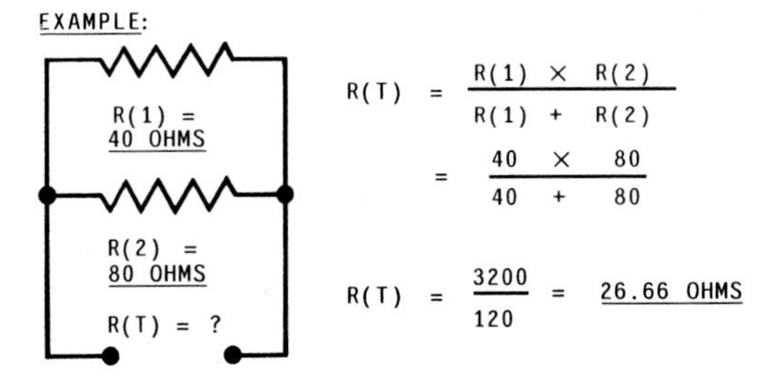

$$R(T) = \frac{R(1) \times R(2)}{R(1) + R(2)}$$

$$= \frac{40 \times 80}{40 + 80}$$

$$R(T) = \frac{3200}{120} = 26.66 \text{ OHMS}$$

COMBINATION CIRCUITS

IN COMBINATION CIRCUITS WE COMBINE SERIES CIRCUITS WITH PARALLEL CIRCUITS. COMBINATION CIRCUITS MAKE IT POSSIBLE TO OBTAIN THE DIFFERENT VOLTAGES OF SERIES CIRCUITS, AND DIFFERENT CURRENTS OF PARALLEL CIRCUITS.

EXAMPLE: 1. PARALLEL-SERIES CIRCUIT:

SOLVE FOR ALL MISSING VALUES.

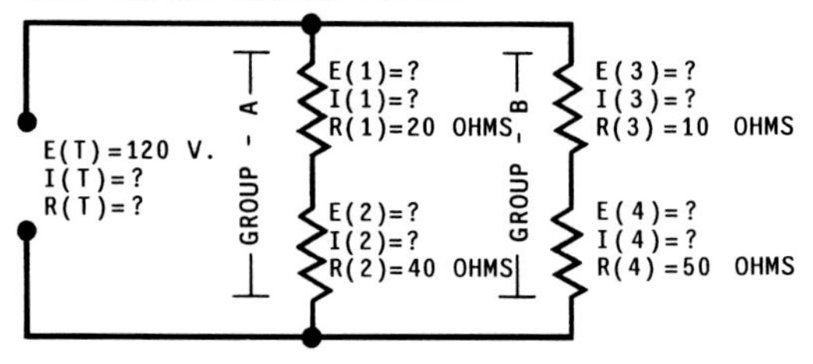

TO SOLVE:

1. FIND THE TOTAL RESISTANCE OF EACH GROUP. BOTH GROUPS ARE SIMPLE SERIES CIRCUITS, SO

   R(1) + R(2) = R(A)
   20 + 40 = 60 OHMS, TOTAL RESISTANCE OF GROUP "A"

   R(3) + R(4) = R(B)
   10 + 50 = 60 OHMS, TOTAL RESISTANCE OF GROUP "B"

2. RE-DRAW THE CIRCUIT, COMBINING RESISTORS (R(1) + R(2)) AND (R(3) + R(4)) SO THAT EACH GROUP WILL HAVE ONLY ONE RESISTOR.

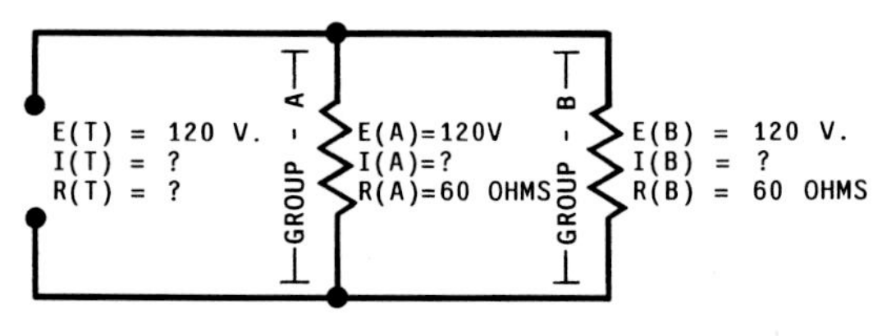

CONTINUED NEXT PAGE

COMBINATION CIRCUITS

NOTE: WE NOW HAVE A SIMPLE PARALLEL CIRCUIT, SO

$$E(T) = E(A) = E(B)$$
$$120\ V = \underline{120\ V} = \underline{120\ V}$$

WE NOW HAVE A PARALLEL CIRCUIT WITH ONLY TWO RESISTORS, AND THEY ARE OF EQUAL VALUE. WE HAVE A CHOICE OF THREE DIFFERENT FORMULAS THAT CAN BE USED TO SOLVE FOR THE TOTAL RESISTANCE OF CIRCUIT.

(1) $$R(T) = \frac{R(A) \times R(B)}{R(A) + R(B)} = \frac{60 \times 60}{60 + 60} = \frac{3600}{120} = \underline{30\ \text{OHMS}}$$

(2) WHEN THE RESISTORS OF A PARALLEL CIRCUIT ARE OF EQUAL VALUE.

$$R(T) = \frac{R}{N} = \frac{60}{2} = \underline{30\ \text{OHMS}}$$

(3) $$\frac{1}{R(T)} = \frac{1}{R(A)} + \frac{1}{R(B)} = \frac{1}{60} + \frac{1}{60} = \frac{2}{60} = \frac{1}{30}$$

$$\frac{1}{R(T)} = \frac{1}{30} \quad \text{OR} \quad 1 \times R(T) = 1 \times 30 \quad \text{OR} \quad R(T) = \underline{30\ \text{OHMS}}$$

3. WE NOW KNOW THE VALUES OF E(T), R(T), E(A), R(A), E(B), R(B), R(1), R(2), R(3), AND R(4). NEXT WE WILL SOLVE FOR I(T), I(A), I(B), I(1), I(2), I(3), AND I(4).

$$\frac{E(T)}{R(T)} = I(T) \text{ OR } \frac{120}{30} = \underline{4} \qquad I(T) = \underline{4\ \text{AMP.}}$$

$$\frac{E(A)}{R(A)} = I(A) \text{ OR } \frac{120}{60} = \underline{2} \qquad I(A) = \underline{2\ \text{AMP.}}$$

$$I(A) = I(1) = I(2) \text{ OR } 2 = \underline{2} = \underline{2} \qquad I(1) = \underline{2\ \text{AMP.}} \quad I(2) = \underline{2\ \text{AMP.}}$$

$$\frac{E(B)}{R(B)} = I(B) \text{ OR } \frac{120}{60} = \underline{2} \qquad I(B) = \underline{2\ \text{AMP.}}$$

$$I(B) = I(3) = I(4) \text{ OR } 2 = \underline{2} = \underline{2} \qquad I(3) = \underline{2\ \text{AMP.}} \quad I(4) = \underline{2\ \text{AMP.}}$$

CONTINUED NEXT PAGE

COMBINATION CIRCUITS

4. WE KNOW THAT RESISTORS #1 and #2 OF GROUP "A" ARE IN SERIES. WE KNOW TOO THAT RESISTORS #3 and #4 OF GROUP "B" ARE IN SERIES. WE HAVE DETERMINED THAT THE TOTAL RESISTANCE OF GROUP "A" IS 2 AMP, AND THE TOTAL RESISTANCE OF GROUP "B" IS 2 AMP. BY USING THE SERIES FORMULA WE CAN SOLVE FOR THE CURRENT VALUE OF EACH RESISTOR.

I(A) = I(1) = I(2)
2 = 2 = 2
I(1) = 2 AMP.
I(2) = 2 AMP.

I(B) = I(3) = I(4)
2 = 2 = 2
I(3) = 2 AMP.
I(4) = 2 AMP.

5. WE WERE GIVEN THE RESISTANCE VALUES OF ALL RESISTORS. R(1) = 20 OHMS, R(2) = 40 OHMS, R(3) = 10 OHMS, AND R(4) = 50 OHMS.

BY USING OHM'S LAW WE CAN DETERMINE THE VOLTAGE DROP ACROSS EACH RESISTOR.

E(1) = R(1) × I(1)
= 20 × 2
E(1) = 40 VOLTS

E(3) = R(3) × I(3)
= 10 × 2
E(3) = 20 VOLTS

E(2) = R(2) × I(2)
= 40 × 2
E(2) = 80 VOLTS

E(4) = R(4) × I(4)
= 50 × 2
E(4) = 100 VOLTS

EXAMPLE: 2. SERIES PARALLEL CIRCUIT:

SOLVE FOR ALL MISSING VALUES

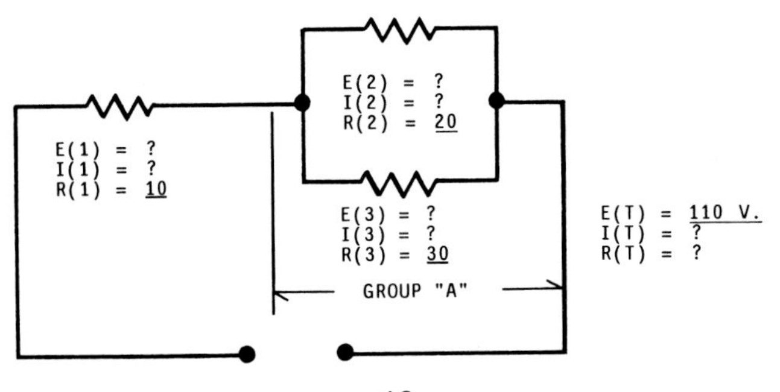

COMBINATION CIRCUITS

TO SOLVE:

1. WE CAN SEE THAT RESISTORS #2 AND #3 ARE IN PARALLEL, AND COMBINED THEY ARE GROUP "A". WHEN THERE ARE ONLY TWO RESISTORS, WE USE THE FOLLOWING FORMULA.

$$R(A) = \frac{R(2) \times R(3)}{R(2) + R(3)} \text{ OR } \frac{20 \times 30}{20 + 30} \text{ OR } \frac{600}{50} \text{ OR } \underline{12 \text{ OHMS}}$$

2. WE CAN NOW RE-DRAW OUR CIRCUIT AS A SIMPLE SERIES CIRCUIT.

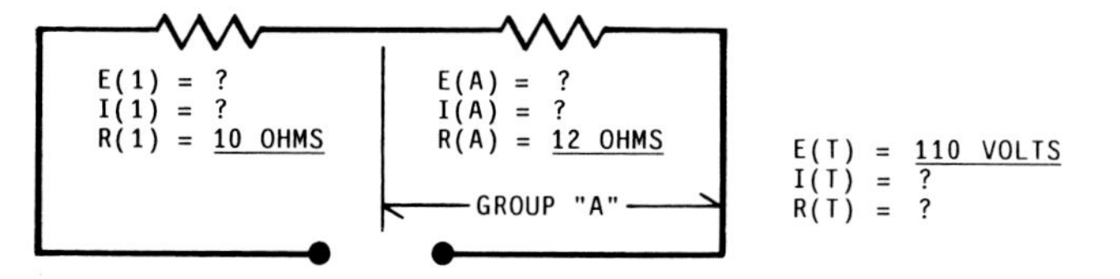

3. IN A SERIES CIRCUIT

R(T) = R(1) + R(A) OR R(T) = 10 + 12 OR 22 OHMS

BY USING OHM'S LAW

$$I(T) = \frac{E(T)}{R(T)} = \frac{110}{22} = \underline{5 \text{ AMP}}$$

IN A SERIES CIRCUIT

I(T) = I(1) = I(A) OR I(T) = 5 AMP, I(1) = 5 AMP AND I(A) = 5 AMP

BY USING OHM'S LAW

E(1) = I(1) × R(1) = 5 × 10 = 50 VOLTS

E(T) - E(1) = E(A) OR 110 - 50 = 60 VOLTS = E(A)

IN A PARALLEL CIRCUIT

E(A) = E(2) = E(3) OR E(A) = 60 VOLTS, E(2) = 60 VOLTS, AND E(3) = 60 VOLTS.

COMBINATION CIRCUITS

BY USING OHM'S LAW

$$I(2) = \frac{E(2)}{R(2)} = \frac{60}{20} = \underline{3\ AMP}$$

$$I(3) = \frac{E(3)}{R(3)} = \frac{60}{30} = \underline{2\ AMP}$$

PROBLEM: SOLVE FOR TOTAL RESISTANCE
RE-DRAW CIRCUIT AS MANY TIMES AS NECESSARY
CORRECT ANSWER IS 100 OHMS

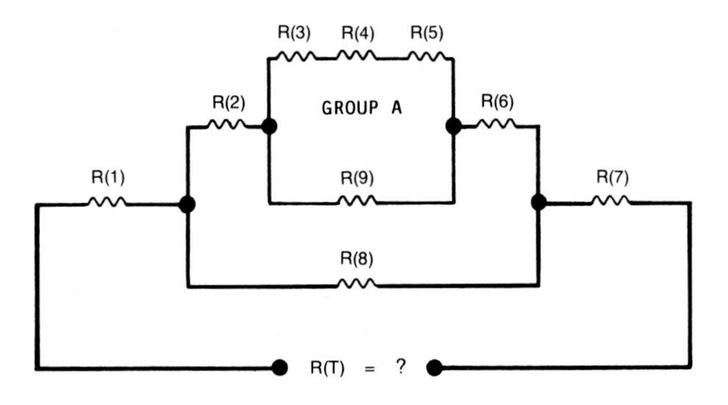

GIVEN VALUES:

R(1) = 15 OHMS

R(2) = 35 OHMS

R(3) = 50 OHMS

R(4) = 40 OHMS

R(5) = 30 OHMS

R(6) = 25 OHMS

R(7) = 10 OHMS

R(8) = 300 OHMS

R(9) = 60 OHMS

ELECTRICAL FORMULAS FOR CALCULATING AMPERES, HORSEPOWER, KILOWATTS, AND KVA

TO FIND	DIRECT CURRENT	ALTERNATING CURRENT		
		SINGLE PHASE	TWO PHASE FOUR WIRE	THREE PHASE
AMPERES WHEN "HP" IS KNOWN	$\frac{HP \times 746}{E \times \%EFF}$	$\frac{HP \times 746}{E \times \%EFF \times PF}$	$\frac{HP \times 746}{E \times \%EFF \times PF \times 2}$	$\frac{HP \times 746}{E \times \%EFF \times PF \times 1.73}$
AMPERES WHEN "KW" IS KNOWN	$\frac{KW \times 1000}{E}$	$\frac{KW \times 1000}{E \times PF}$	$\frac{KW \times 1000}{E \times PF \times 2}$	$\frac{KW \times 1000}{E \times PF \times 1.73}$
AMPERES WHEN "KVA" IS KNOWN		$\frac{KVA \times 1000}{E}$	$\frac{KVA \times 1000}{E \times 2}$	$\frac{KVA \times 1000}{E \times 1.73}$
KILOWATTS	$\frac{E \times I}{1000}$	$\frac{E \times I \times PF}{1000}$	$\frac{E \times I \times PF \times 2}{1000}$	$\frac{E \times I \times PF \times 1.73}{1000}$
KILOVOLT-AMPERES "KVA"		$\frac{E \times I}{1000}$	$\frac{E \times I \times 2}{1000}$	$\frac{E \times I \times 1.73}{1000}$
HORSEPOWER	$\frac{E \times I \times \%EFF}{746}$	$\frac{E \times I \times \%EFF \times PF}{746}$	$\frac{E \times I \times \%EFF \times PF \times 2}{746}$	$\frac{E \times I \times \%EFF \times PF \times 1.73}{746}$

PERCENT EFFICIENCY = $\%EFF = \frac{OUTPUT}{INPUT}$ POWER FACTOR = $PF = \frac{\text{POWER USED (WATTS)}}{\text{APPARENT POWER}} = \frac{KW}{KVA}$

NOTE: DIRECT CURRENT FORMULAS DO NOT USE (PF, 2, OR 1.73)
SINGLE PHASE FORMULAS DO NOT USE (2 OR 1.73)
TWO PHASE-FOUR WIRE FORMULAS DO NOT USE (1.73)
THREE PHASE FORMULAS DO NOT USE (2)

TO FIND AMPERES

DIRECT CURRENT:

A. WHEN HORSEPOWER IS KNOWN:

$$\text{AMPERES} = \frac{\text{HORSEPOWER} \times 746}{\text{VOLTS} \times \text{EFFICIENCY}} \quad \text{OR} \quad I = \frac{HP \times 746}{E \times \%EFF}$$

WHAT CURRENT WILL A TRAVEL-TRAILER TOILET DRAW WHEN EQUIPPED WITH A 12 VOLT, 1/8 HP MOTOR, HAVING A 96% EFFICIENCY RATING?

$$I = \frac{HP \times 746}{E \times \%EFF} = \frac{746 \times 1/8}{12 \times 0.96} = \frac{93.25}{11.52} = \underline{8.09 \text{ AMP}}$$

B. WHEN KILOWATTS ARE KNOWN:

$$\text{AMPERES} = \frac{\text{KILOWATTS} \times 1000}{\text{VOLTS}} \quad \text{OR} \quad I = \frac{KW \times 1000}{E}$$

A 75 KW, 240 VOLT, DIRECT CURRENT GENERATOR IS USED TO POWER A VARIABLE-SPEED CONVEYOR BELT AT A ROCK CRUSHING PLANT. DETERMINE THE CURRENT.

$$I = \frac{KW \times 1000}{E} = \frac{75 \times 1000}{240} = \underline{312.5 \text{ AMPERES}}$$

SINGLE PHASE:

A. WHEN WATTS, VOLTS, AND POWER-FACTOR ARE KNOWN:

$$\text{AMPERES} = \frac{\text{WATTS}}{\text{VOLTS} \times \text{POWER-FACTOR}}$$

OR

$$I = \frac{P}{E \times PF}$$

DETERMINE THE CURRENT WHEN A CIRCUIT HAS A 1500 WATT LOAD, A POWER-FACTOR OF 86%, AND OPERATES FROM A SINGLE-PHASE 230 VOLT SOURCE.

$$I = \frac{1500}{230 \times 0.86} = \frac{1500}{197.8} = \underline{7.58 \text{ AMP}}$$

TO FIND AMPERES

SINGLE PHASE:

B. WHEN HORSEPOWER IS KNOWN:

$$\text{AMPERES} = \frac{\text{HORSEPOWER} \times 746}{\text{VOLTS} \times \text{EFFICIENCY} \times \text{POWER-FACTOR}}$$

DETERMINE THE AMP-LOAD OF A SINGLE-PHASE, 1/2 HP, 115 VOLT MOTOR. THE MOTOR HAS AN EFFICIENCY RATING OF 92%, AND A POWER-FACTOR OF 80%.

$$I = \frac{HP \times 746}{E \times \%EFF \times PF} = \frac{1/2 \times 746}{115 \times 0.92 \times 0.80} = \frac{373}{84.64}$$

$$I = \underline{4.4\ \text{AMP}}$$

C. WHEN KILOWATTS ARE KNOWN:

$$\text{AMPERES} = \frac{\text{KILOWATTS} \times 1000}{\text{VOLTS} \times \text{POWER-FACTOR}} \quad \text{OR} \quad I = \frac{KW \times 1000}{E \times PF}$$

A 230 VOLT SINGLE PHASE CIRCUIT HAS A 12 KW POWER LOAD, AND OPERATES AT 84% POWER-FACTOR. DETERMINE THE CURRENT.

$$I = \frac{KW \times 1000}{E \times PF} = \frac{12 \times 1000}{230 \times 0.84} = \frac{12{,}000}{193.2} = \underline{62\ \text{AMP}}$$

D. WHEN KILOVOLT-AMPERE IS KNOWN:

$$\text{AMPERES} = \frac{\text{KILOVOLT-AMPERE} \times 1000}{\text{VOLTS}} \quad \text{OR} \quad I = \frac{KVA \times 1000}{E}$$

A 115 VOLT, 2 KVA, SINGLE PHASE GENERATOR OPERATING AT FULL LOAD WILL DELIVER 17.4 AMPERES. (PROVE)

$$I = \frac{2 \times 1000}{115} = \frac{2000}{115} = \underline{17.4\ \text{AMP}}$$

REMEMBER: BY DEFINITION AMPERES IS THE RATE OF THE FLOW OF THE CURRENT.

TO FIND AMPERES

TWO-PHASE, FOUR WIRE:

NOTE: FOR THREE WIRE, TWO-PHASE CIRCUITS, THE CURRENT IN THE COMMON CONDUCTOR IS 1.41 GREATER THAN IN EITHER OF THE OTHER TWO CONDUCTORS.

A. WHEN WATTS, VOLTS, AND POWER-FACTOR ARE KNOWN:

$$\text{AMPERES} = \frac{\text{WATTS}}{\text{VOLTS} \times \text{POWER-FACTOR} \times 2} = \frac{P}{E \times PF \times 2}$$

DETERMINE THE CURRENT WHEN A CIRCUIT HAS A 1500 WATT LOAD, A POWER-FACTOR OF 86%, AND OPERATES FROM A TWO PHASE, 230 VOLT SOURCE.

$$I = \frac{P}{E \times PF \times 2} = \frac{1500}{230 \times 0.86 \times 2} = \frac{1500}{395.6}$$

$$I = \underline{3.79 \text{ AMP}}$$

B. WHEN HORSEPOWER IS KNOWN:

$$\text{AMPERES} = \frac{\text{HORSEPOWER} \times 746}{\text{VOLTS} \times \text{EFFICIENCY} \times \text{POWER-FACTOR} \times 2}$$

OR

$$I = \frac{HP \times 746}{E \times \%EFF \times PF \times 2}$$

DETERMINE THE AMP-LOAD OF A TWO-PHASE, 1/2 HP, 230 VOLT MOTOR. THE MOTOR HAS AN EFFICIENCY RATING OF 92%, AND A POWER-FACTOR OF 80%.

$$I = \frac{HP \times 746}{E \times \%EFF \times PF \times 2} = \frac{1/2 \times 746}{230 \times 0.92 \times 0.80 \times 2}$$

$$= \frac{373}{339} = \underline{1.1 \text{ AMP}}$$

NOTE:

$$\frac{\text{CONSUMED POWER}}{\text{APPARENT POWER}} = \frac{KW}{KVA} = \text{POWER-FACTOR (PF)}$$

TO FIND AMPERES

TWO-PHASE, FOUR WIRE:

C. WHEN KILOWATTS ARE KNOWN:

$$\text{AMPERES} = \frac{\text{KILOWATTS} \times 1000}{\text{VOLTS} \times \text{POWER-FACTOR} \times 2}$$

OR

$$I = \frac{KW \times 1000}{E \times PF \times 2}$$

A 230 VOLT, TWO-PHASE CIRCUIT, HAS A 12 KW POWER LOAD, AND OPERATES AT 84% POWER-FACTOR. DETERMINE THE CURRENT.

$$I = \frac{KW \times 1000}{E \times PF \times 2} = \frac{12 \times 1000}{230 \times 0.84 \times 2} = \frac{12{,}000}{386.4}$$

$$= \underline{31 \text{ AMP}}$$

D. WHEN KILOVOLT-AMPERE IS KNOWN:

$$\text{AMPERES} = \frac{\text{KILOVOLT-AMPERE} \times 1000}{\text{VOLTS} \times 2}$$

OR

$$I = \frac{KVA \times 1000}{E \times 2}$$

A 230 VOLT, 4 KVA, TWO-PHASE GENERATOR OPERATING AT FULL LOAD WILL DELIVER 8.7 AMPERES. (PROVE)

$$I = \frac{4 \times 1000}{230 \times 2} = \frac{4000}{460} = \underline{8.7 \text{ AMP}}$$

TO FIND AMPERES

THREE-PHASE:

A. WHEN WATTS, VOLTS, AND POWER-FACTOR ARE KNOWN:

$$\text{AMPERES} = \frac{\text{WATTS}}{\text{VOLTS} \times \text{POWER-FACTOR} \times 1.73}$$

OR

$$I = \frac{P}{E \times PF \times 1.73}$$

DETERMINE THE CURRENT WHEN A CIRCUIT HAS A 1500 WATT LOAD, A POWER-FACTOR OF 86%, AND OPERATES FROM A THREE-PHASE, 230 VOLT SOURCE.

$$I = \frac{P}{E \times PF \times 1.73} = \frac{1500}{230 \times 0.86 \times 1.73} = \frac{1500}{342.2}$$

$$= \underline{4.4 \text{ AMP}}$$

B. WHEN HORSEPOWER IS KNOWN:

$$\text{AMPERES} = \frac{\text{HORSEPOWER} \times 746}{\text{VOLTS} \times \text{EFFICIENCY} \times \text{POWER-FACTOR} \times 1.73}$$

OR

$$I = \frac{HP \times 746}{E \times \%EFF \times PF \times 1.73}$$

DETERMINE THE AMP-LOAD OF A THREE-PHASE, 1/2 HP, 230 VOLT MOTOR. THE MOTOR HAS AN EFFICIENCY RATING OF 92%, AND A POWER-FACTOR OF 80%.

$$I = \frac{HP \times 746}{E \times \%EFF \times PF \times 1.73} = \frac{1/2 \times 746}{230 \times .92 \times .80 \times 1.73}$$

$$= \frac{373}{293} = \underline{1.27 \text{ AMP}}$$

TO FIND AMPERES

THREE-PHASE:

C. WHEN KILOWATTS ARE KNOWN:

$$\text{AMPERES} = \frac{\text{KILOWATTS} \times 1000}{\text{VOLTS} \times \text{POWER-FACTOR} \times 1.73}$$

OR

$$I = \frac{KW \times 1000}{E \times PF \times 1.73}$$

A 230 VOLT, THREE-PHASE CIRCUIT, HAS A 12 KW POWER LOAD, AND OPERATES AT 84% POWER-FACTOR. DETERMINE THE CURRENT.

$$I = \frac{KW \times 1000}{E \times PF \times 1.73} = \frac{12{,}000}{230 \times 0.84 \times 1.73} = \frac{12{,}000}{334.24}$$

$I = \underline{36 \text{ AMP}}$

D. WHEN KILOVOLT-AMPERE IS KNOWN:

$$\text{AMPERES} = \frac{\text{KILOVOLT-AMPERE} \times 1000}{E \times 1.73} = \frac{KVA \times 1000}{E \times 1.73}$$

A 230 VOLT, 4 KVA, THREE PHASE GENERATOR OPERATING AT FULL LOAD WILL DELIVER 10 AMPERES. (PROVE)

$$I = \frac{KVA \times 1000}{E \times 1.73} = \frac{4 \times 1000}{230 \times 1.73} = \frac{4000}{397.9}$$

$I = \underline{10 \text{ AMP}}$

NOTE: TO BETTER UNDERSTAND THE PRECEDING FORMULAS:

1. TWO-PHASE CURRENT × 2 = SINGLE-PHASE CURRENT.
2. THREE-PHASE CURRENT × 1.73 = SINGLE PHASE CURRENT.
3. THE CURRENT IN THE COMMON CONDUCTOR OF A TWO-PHASE (THREE WIRE) CIRCUIT IS 141% GREATER THAN EITHER OF THE OTHER TWO CONDUCTORS OF THAT CIRCUIT.

TO FIND HORSEPOWER

DIRECT CURRENT:

$$\text{HORSEPOWER} = \frac{\text{VOLTS} \times \text{AMPERES} \times \text{EFFICIENCY}}{746}$$

A 12 VOLT MOTOR DRAWS A CURRENT OF 8.09 AMPERES, AND HAS AN EFFICIENCY RATING OF 96%. DETERMINE THE HORSEPOWER.

$$\text{HP} = \frac{\text{E} \times \text{I} \times \text{\%EFF}}{746} = \frac{12 \times 8.09 \times 0.96}{746} = \frac{93.19}{746}$$

$$= 0.1249 = \underline{1/8 \text{ HP}}$$

SINGLE-PHASE:

$$\text{HP} = \frac{\text{VOLTS} \times \text{AMPERES} \times \text{EFFICIENCY} \times \text{POWER-FACTOR}}{746}$$

A SINGLE-PHASE, 115 VOLT (AC) MOTOR HAS AN EFFICIENCY RATING OF 92%, AND A POWER-FACTOR OF 80%. DETERMINE THE HORSEPOWER IF THE AMP-LOAD IS 4.4 AMPERES.

$$\text{HP} = \frac{\text{E} \times \text{I} \times \text{\%EFF} \times \text{PF}}{746} = \frac{115 \times 4.4 \times 0.92 \times 0.80}{746}$$

$$= \frac{372.416}{746} = 0.4992 = \underline{1/2 \text{ HP}}$$

TWO-PHASE:

$$\text{HP} = \frac{\text{VOLTS} \times \text{AMPERES} \times \text{EFFICIENCY} \times \text{POWER-FACTOR} \times 2}{746}$$

DETERMINE THE HORSEPOWER OF A TWO-PHASE, 230 VOLT (AC) MOTOR. THE MOTOR HAS AN EFFICIENCY RATING OF 92%, A POWER-FACTOR OF 80%, AND AN AMP-LOAD OF 1.1 AMPERES.

$$\text{HP} = \frac{\text{E} \times \text{I} \times \text{\%EFF} \times \text{PF} \times 2}{746} = \frac{230 \times 1.1 \times .92 \times .8 \times 2}{746}$$

$$= \frac{372.416}{746} = 0.4992 = \underline{1/2 \text{ HP}}$$

TO FIND HORSEPOWER

THREE-PHASE:

$$HP = \frac{\text{VOLTS} \times \text{AMPERES} \times \text{EFFICIENCY} \times \text{POWER-FACTOR} \times 1.73}{746}$$

A THREE-PHASE, 460 VOLT MOTOR DRAWS A CURRENT OF 52 AMPERES. THE MOTOR HAS AN EFFICIENCY RATING OF 94%, AND A POWER FACTOR OF 80%. DETERMINE THE HORSEPOWER.

$$HP = \frac{E \times I \times \%EFF \times PF \times 1.73}{746}$$

$$= \frac{460 \times 52 \times 0.94 \times 0.80 \times 1.73}{746}$$

$$= \underline{41.7 \text{ HP}}$$

TO FIND WATTS

THE ELECTRICAL POWER IN ANY PART OF A CIRCUIT IS EQUAL TO THE CURRENT IN THAT PART MULTIPLIED BY THE VOLTAGE ACROSS THAT PART OF THE CIRCUIT.

A WATT IS THE POWER USED WHEN ONE VOLT CAUSES ONE AMPERE TO FLOW IN A CIRCUIT.

ONE HORSEPOWER IS THE AMOUNT OF ENERGY REQUIRED TO LIFT 33,000 POUNDS, ONE FOOT, IN ONE MINUTE. THE ELECTRICAL EQUIVALENT OF ONE HORSEPOWER IS 745.6 WATTS. ONE WATT IS THE AMOUNT OF ENERGY REQUIRED TO LIFT 44.26 POUNDS, ONE FOOT, IN ONE MINUTE. WATTS IS POWER, AND POWER IS THE AMOUNT OF WORK DONE IN A GIVEN TIME.

1. **WHEN VOLTS AND AMPERES ARE KNOWN:**

A. POWER (WATTS) = VOLTS × AMPERES

A 120 VOLT A-C CIRCUIT DRAWS A CURRENT OF 5 AMPERES: DETERMINE THE POWER CONSUMPTION.

$P = E \times I = 120 \times 5 = \underline{600 \text{ WATTS}}$

WE CAN NOW DETERMINE THE RESISTANCE OF THIS CIRCUIT.

(1.) $\text{POWER} = \text{RESISTANCE} \times (\text{AMPERES})^2$

$P = R \times (I)^2$ OR $600 = R \times 25$

$\frac{600}{25} = R$ OR $R = \underline{24 \text{ OHMS}}$

(2.) $\text{POWER} = \frac{(\text{VOLTS})^2}{\text{RESISTANCE}}$ OR $P = \frac{(E)^2}{R}$

$R \times 600 = (120)^2$ OR $R = \frac{14,400}{600}$

$R = \underline{24 \text{ OHMS}}$

NOTE: REFER TO FORMULAS OF THE OHM'S LAW CHART ON PAGE 1.

TO FIND KILOWATTS

DIRECT CURRENT:

$$\text{KILOWATTS} = \frac{\text{VOLTS} \times \text{AMPERES}}{1000}$$

A 120 VOLT (DC) MOTOR DRAWS A CURRENT OF 40 AMPERES. DETERMINE THE KILOWATTS.

$$\text{KW} = \frac{E \times I}{1000} = \frac{120 \times 40}{1000} = \frac{4800}{1000} = \underline{4.8 \text{ KW}}$$

SINGLE-PHASE:

$$\text{KILOWATTS} = \frac{\text{VOLTS} \times \text{AMPERES} \times \text{POWER-FACTOR}}{1000}$$

A SINGLE-PHASE, 115 VOLT (AC) MOTOR DRAWS A CURRENT OF 20 AMPERES, AND HAS A POWER-FACTOR RATING OF 86%. DETERMINE THE KILOWATTS.

$$\text{KW} = \frac{E \times I \times PF}{1000} = \frac{115 \times 20 \times 0.86}{1000} = \frac{1978}{1000}$$

$$= 1.978 = \underline{2 \text{ KW}}$$

TWO-PHASE:

$$\text{KILOWATTS} = \frac{\text{VOLTS} \times \text{AMPERES} \times \text{POWER-FACTOR} \times 2}{1000}$$

A TWO-PHASE, 230 VOLT (AC) MOTOR WITH A POWER-FACTOR OF 92%, DRAWS A CURRENT OF 55 AMPERES. DETERMINE THE KILOWATTS.

$$\text{KW} = \frac{E \times I \times PF \times 2}{1000} = \frac{230 \times 55 \times 0.92 \times 2}{1000}$$

$$= \frac{23,276}{1000} = 23.276 = \underline{23 \text{ KW}}$$

TO FIND KILOWATTS

THREE-PHASE:

$$\text{KILOWATTS} = \frac{\text{VOLTS} \times \text{AMPERES} \times \text{POWER-FACTOR} \times 1.73}{1000}$$

A THREE-PHASE, 460 VOLT MOTOR DRAWS A CURRENT OF 52 AMPERES, AND HAS A POWER-FACTOR RATED AT 80%. DETERMINE THE KILOWATTS.

$$KW = \frac{E \times I \times PF \times 1.73}{1000} = \frac{460 \times 52 \times 0.80 \times 1.73}{1000}$$

$$= \frac{33,105}{1000} = 33.105 = \underline{33\ KW}$$

KIRCHHOFF'S LAWS

FIRST LAW (CURRENT)

THE SUM OF THE CURRENTS ARRIVING AT ANY POINT IN A CIRCUIT MUST EQUAL THE SUM OF THE CURRENTS LEAVING THAT POINT.

SECOND LAW (VOLTAGE)

THE TOTAL VOLTAGE APPLIED TO ANY CLOSED CIRCUIT PATH IS ALWAYS EQUAL TO THE SUM OF THE VOLTAGE DROPS IN THAT PATH.

OR

THE ALGEBRAIC SUM OF ALL THE VOLTAGES ENCOUNTERED IN ANY LOOP EQUALS ZERO.

TO FIND KILOVOLT-AMPERES

SINGLE-PHASE:

$$\text{KILOVOLT-AMPERES} = \frac{\text{VOLTS} \times \text{AMPERES}}{1000}$$

A SINGLE-PHASE, 240 VOLT GENERATOR DELIVERS 41.66 AMPERES AT FULL LOAD. DETERMINE THE KILOVOLT-AMPERES RATING.

$$\text{KVA} = \frac{E \times I}{1000} = \frac{240 \times 41.66}{1000} = \frac{10,000}{1000} = \underline{10 \text{ KVA}}$$

TWO-PHASE:

$$\text{KILOVOLT-AMPERES} = \frac{\text{VOLTS} \times \text{AMPERES} \times 2}{1000}$$

A TWO-PHASE, 230 VOLT GENERATOR DELIVERS 55 AMPERES. DETERMINE THE KILOVOLT-AMPERES RATING.

$$\text{KVA} = \frac{E \times I \times 2}{1000} = \frac{230 \times 55 \times 2}{1000} = \frac{25,300}{1000}$$

$$= 25.3 = \underline{25 \text{ KVA}}$$

THREE-PHASE:

$$\text{KILOVOLT-AMPERES} = \frac{\text{VOLTS} \times \text{AMPERES} \times 1.73}{1000}$$

A THREE-PHASE, 460 VOLT GENERATOR DELIVERS 52 AMPERES. DETERMINE THE KILOVOLT-AMPERES RATING.

$$\text{KVA} = \frac{E \times I \times 1.73}{1000} = \frac{460 \times 52 \times 1.73}{1000} = \frac{41,382}{1000}$$

$$= 41.382 = \underline{41 \text{ KVA}}$$

NOTE: KVA = APPARENT POWER = POWER BEFORE USED, SUCH AS THE RATING OF A TRANSFORMER.

TO FIND

CAPACITANCE (C):

$$C = \frac{Q}{E} \quad \text{OR CAPACITANCE} = \frac{\text{COULOMBS}}{\text{VOLTS}}$$

CAPACITANCE IS THE PROPERTY OF A CIRCUIT OR BODY THAT PERMITS IT TO STORE AN ELECTRICAL CHARGE EQUAL TO THE ACCUMULATED CHARGE DIVIDED BY THE VOLTAGE. EXPRESSED IN FARADS.

A. TO DETERMINE THE TOTAL CAPACITY OF CAPACITORS, AND / OR CONDENSERS CONNECTED IN SERIES.

$$\frac{1}{C(T)} = \frac{1}{C(1)} + \frac{1}{C(2)} + \frac{1}{C(3)} + \frac{1}{C(4)}$$

DETERMINE THE TOTAL CAPACITY OF FOUR EACH, 12 MICROFARAD CAPACITORS CONNECTED IN SERIES.

$$\frac{1}{C(T)} = \frac{1}{C(1)} + \frac{1}{C(2)} + \frac{1}{C(3)} + \frac{1}{C(4)}$$

$$= \frac{1}{12} + \frac{1}{12} + \frac{1}{12} + \frac{1}{12}$$

$$\frac{1}{C(T)} = \frac{4}{12} \quad \text{OR} \quad C(T) \times 4 = 12 \quad \text{OR} \quad C(T) = \frac{12}{4}$$

C(T) = <u>3 MICROFARADS</u>

B. TO DETERMINE THE TOTAL CAPACITY OF CAPACITORS, AND / OR CONDENSERS CONNECTED IN PARALLEL.

$$C(T) = C(1) + C(2) + C(3) + C(4)$$

DETERMINE THE TOTAL CAPACITY OF FOUR EACH, 12 MICROFARAD CAPACITORS CONNECTED IN PARALLEL.

$$C(T) = C(1) + C(2) + C(3) + C(4)$$
$$C(T) = 12 + 12 + 12 + 12$$

C(T) = <u>48 MICROFARADS</u>

A FARAD IS THE UNIT OF CAPACITANCE OF A CONDENSER THAT RETAINS ONE COULOMB OF CHARGE WITH ONE VOLT DIFFERENCE OF POTENTIAL.

1 FARAD = 1,000,000 MICROFARADS

6-DOT COLOR CODE FOR MICA AND MOLDED PAPER CAPACITORS

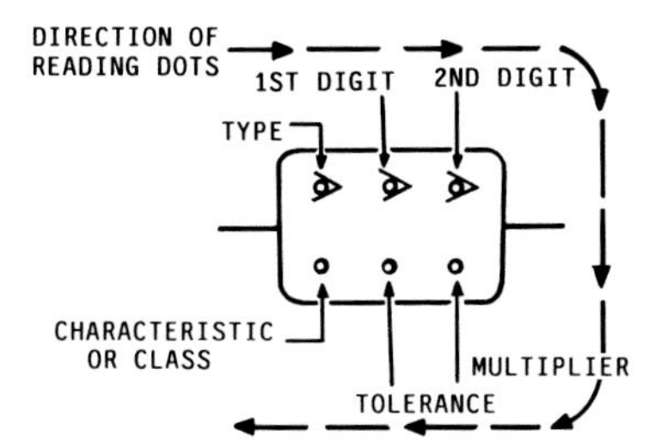

TYPE	COLOR	1ST DIGIT	2ND DIGIT	MULTIPLIER	TOLERANCE (%)	CHARACTERISTIC OR CLASS
JAN, MICA	BLACK	0	0	1	± 1	APPLIES TO TEMPERATURE COEFFICIENT OR METHODS OF TESTING
	BROWN	1	1	10	± 2	
	RED	2	2	100	± 3	
	ORANGE	3	3	1,000	± 4	
	YELLOW	4	4	10,000	± 5	
	GREEN	5	5	100,000	± 6	
	BLUE	6	6	1,000,000	± 7	
	VIOLET	7	7	10,000,000	± 8	
	GRAY	8	8	100,000,000	± 9	
EIA, MICA	WHITE	9	9	1,000,000,000		
	GOLD			.1	±10	
MOLDED PAPER	SILVER			.01	±20	
	BODY					

MAXIMUM PERMISSIBLE CAPACITOR KVAR FOR USE WITH

OPEN-TYPE THREE-PHASE SIXTY-CYCLE INDUCTION MOTORS

MOTOR RATING HP	3600 RPM		1800 RPM		1200 RPM	
	MAXIMUM CAPACITOR RATING KVAR	REDUCTION IN LINE CURRENT %	MAXIMUM CAPACITOR RATING KVAR	REDUCTION IN LINE CURRENT %	MAXIMUM CAPACITOR RATING KVAR	REDUCTION IN LINE CURRENT %
10	3	10	3	11	3.5	14
15	4	9	4	10	5	13
20	5	9	5	10	6.5	12
25	6	9	6	10	7.5	11
30	7	8	7	9	9	11
40	9	8	9	9	11	10
50	12	8	11	9	13	10
60	14	8	14	8	15	10
75	17	8	16	8	18	10
100	22	8	21	8	25	9
125	27	8	26	8	30	9
150	32.5	8	30	8	35	9
200	40	8	37.5	8	42.5	9

	900 RPM		720 RPM		600 RPM	
10	5	21	6.5	27	7.5	31
15	6.5	18	8	23	9.5	27
20	7.5	16	9	21	12	25
25	9	15	11	20	14	23
30	10	14	12	18	16	22
40	12	13	15	16	20	20
50	15	12	19	15	24	19
60	18	11	22	15	27	19
75	21	10	26	14	32.5	18
100	27	10	32.5	13	40	17
125	32.5	10	40	13	47.5	16
150	37.5	10	47.5	12	52.5	15
200	47.5	10	60	12	65	14

NOTE: IF CAPACITORS OF A LOWER RATING THAN THE VALUES GIVEN IN THE TABLE ARE USED, THE PERCENTAGE REDUCTION IN LINE CURRENT GIVEN IN THE TABLE SHOULD BE REDUCED PROPORTIONALLY.

POWER-FACTOR CORRECTION

TABLE VALUES × KW LOAD = KVA OF CAPACITORS NEEDED TO CORRECT FROM EXISTING TO DESIRED POWER FACTOR.

EXISTING POWER FACTOR %	CORRECTED POWER FACTOR					
	100%	95%	90%	85%	80%	75%
50	1.732	1.403	1.247	1.112	0.982	0.850
52	1.643	1.314	1.158	1.023	0.893	0.761
54	1.558	1.229	1.073	0.938	0.808	0.676
55	1.518	1.189	1.033	0.898	0.768	0.636
56	1.479	1.150	0.994	0.859	0.729	0.597
58	1.404	1.075	0.919	0.784	0.654	0.522
60	1.333	1.004	0.848	0.713	0.583	0.451
62	1.265	0.936	0.780	0.645	0.515	0.383
64	1.201	0.872	0.716	0.581	0.451	0.319
65	1.168	0.839	0.683	0.548	0.418	0.286
66	1.139	0.810	0.654	0.519	0.389	0.257
68	1.078	0.749	0.593	0.458	0.328	0.196
70	1.020	0.691	0.535	0.400	0.270	0.138
72	0.964	0.635	0.479	0.344	0.214	0.082
74	0.909	0.580	0.424	0.289	0.159	0.027
75	0.882	0.553	0.397	0.262	0.132	
76	0.855	0.526	0.370	0.235	0.105	
78	0.802	0.473	0.317	0.182	0.052	
80	0.750	0.421	0.265	0.130		
82	0.698	0.369	0.213	0.078		
84	0.646	0.317	0.161			
85	0.620	0.291	0.135			
86	0.594	0.265	0.109			
88	0.540	0.211	0.055			
90	0.485	0.156				
92	0.426	0.097				
94	0.363	0.034				
95	0.329					

TYPICAL PROBLEM: WITH A LOAD OF 500 KW AT 70% POWER FACTOR, IT IS DESIRED TO FIND THE KVA OF CAPACITORS REQUIRED TO CORRECT THE POWER FACTOR TO 85%.

SOLUTION: FROM THE TABLE SELECT THE MULTIPLYING FACTOR 0.400 CORRESPONDING TO THE EXISTING 70%, AND THE CORRECTED 85% POWER FACTOR. 0.400 × 500 = 200 KVA OF CAPACITORS REQUIRED.

TO FIND

INDUCTION (L):

INDUCTION IS THE PRODUCTION OF MAGNETIZATION OF ELECTRIFICATION IN A BODY BY THE PROXIMITY OF A MAGNETIC FIELD OR ELECTRIC CHARGE, OR OF THE ELECTRIC CURRENT IN A CONDUCTOR BY THE VARIATION OF THE MAGNETIC FIELD IN ITS VICINITY. EXPRESSED IN HENRYS.

A. TO FIND THE TOTAL INDUCTION OF COILS CONNECTED IN SERIES.

$$L(T) = L(1) + L(2) + L(3) + L(4)$$

DETERMINE THE TOTAL INDUCTION OF FOUR COILS CONNECTED IN SERIES. EACH COIL HAS AN INDUCTANCE VALUE OF FOUR HENRYS.

$$\begin{aligned} L(T) &= L(1) + L(2) + L(3) + L(4) \\ &= 4 + 4 + 4 + 4 \\ &= \underline{16 \text{ HENRYS}} \end{aligned}$$

B. TO FIND THE TOTAL INDUCTION OF COILS CONNECTED IN PARALLEL.

$$\frac{1}{L(T)} = \frac{1}{L(1)} + \frac{1}{L(2)} + \frac{1}{L(3)} + \frac{1}{L(4)}$$

DETERMINE THE TOTAL INDUCTION OF FOUR COILS CONNECTED IN PARALLEL. EACH COIL HAS AN INDUCTANCE VALUE OF FOUR HENRYS.

$$\begin{aligned} \frac{1}{L(T)} &= \frac{1}{L(1)} + \frac{1}{L(2)} + \frac{1}{L(3)} + \frac{1}{L(4)} \\ &= \frac{1}{4} + \frac{1}{4} + \frac{1}{4} + \frac{1}{4} \end{aligned}$$

$$\frac{1}{L(T)} = \frac{4}{4} \quad \text{OR} \quad L(T) \times 4 = 1 \times 4 \quad \text{OR} \quad L(T) = \frac{4}{4}$$

$$L(T) = \underline{1 \text{ HENRY}}$$

AN INDUCTION COIL IS A DEVICE, CONSISTING OF TWO CONCENTRIC COILS AND AN INTERRUPTER, THAT CHANGES A LOW STEADY VOLTAGE INTO A HIGH INTERMITTENT ALTERNATING VOLTAGE BY ELECTROMAGNETIC INDUCTION, MOST OFTEN USED AS A SPARK COIL.

TO FIND

IMPEDANCE (Z):

IMPEDANCE IS THE TOTAL OPPOSITION TO AN ALTERNATING CURRENT PRESENTED BY A CIRCUIT. EXPRESSED IN OHMS.

A. WHEN VOLTS AND AMPERES ARE KNOWN:

$$\text{IMPEDANCE} = \frac{\text{VOLTS}}{\text{AMPERES}} \quad \text{OR} \quad Z = \frac{E}{I}$$

DETERMINE THE IMPEDANCE OF A 120 VOLT A-C CIRCUIT THAT DRAWS A CURRENT OF FOUR AMPERES.

$$Z = \frac{E}{I} = \frac{120}{4} = \underline{30 \text{ OHMS}}$$

B. WHEN RESISTANCE AND REACTANCE ARE KNOWN:

$$Z = \sqrt{\text{RESISTANCE}^2 + \text{REACTANCE}^2}$$

$$= \sqrt{R^2 + X^2}$$

DETERMINE THE IMPEDANCE OF AN A-C CIRCUIT WHEN THE RESISTANCE IS 6 OHMS, AND THE REACTANCE IS 8 OHMS.

$$Z = \sqrt{R^2 + X^2} = \sqrt{36 + 64} = \sqrt{100}$$

$$= \underline{10 \text{ OHMS}}$$

C. WHEN RESISTANCE, INDUCTIVE REACTANCE, AND CAPACITIVE REACTANCE ARE KNOWN:

$$Z = \sqrt{R^2 + (X(L) - X(C))^2}$$

DETERMINE THE IMPEDANCE OF AN A-C CIRCUIT WHICH HAS A RESISTANCE OF 6 OHMS, AN INDUCTIVE REACTANCE OF 18 OHMS, AND A CAPACITIVE REACTANCE OF 10 OHMS.

$$Z = \sqrt{R^2 + (X(L) - X(C))^2}$$

$$= \sqrt{6^2 + (18 - 10)^2} = \sqrt{6^2 + (8)^2}$$

$$= \sqrt{36 + 64} = \sqrt{100} = \underline{10 \text{ OHMS}}$$

TO FIND

REACTANCE (X):

REACTANCE IN A CIRCUIT IS THE OPPOSITION TO AN ALTERNATING CURRENT CAUSED BY INDUCTANCE AND CAPACITANCE, EQUAL TO THE DIFFERENCE BETWEEN CAPACITIVE AND INDUCTIVE REACTANCE. EXPRESSED IN OHMS.

A. INDUCTIVE REACTANCE X(L)

INDUCTIVE REACTANCE IS THAT ELEMENT OF REACTANCE IN A CIRCUIT CAUSED BY SELF-INDUCTANCE.

$$X(L) = 2 \times 3.1416 \times \text{FREQUENCY} \times \text{INDUCTANCE}$$

$$= 6.28 \times F \times L$$

DETERMINE THE REACTANCE OF A FOUR-HENRY COIL ON A 60 CYCLE, A-C CIRCUIT.

$$X(L) = 6.28 \times F \times L = 6.28 \times 60 \times 4$$

$$= \underline{1507 \text{ OHMS}}$$

B. CAPACITIVE REACTANCE X(C)

CAPACITIVE REACTANCE IS THAT ELEMENT OF REACTANCE IN A CIRCUIT CAUSED BY CAPACITANCE.

$$X(C) = \frac{1}{2 \times 3.1416 \times \text{FREQUENCY} \times \text{CAPACITANCE}}$$

$$= \frac{1}{6.28 \times F \times C}$$

DETERMINE THE REACTANCE OF A FOUR MICROFARAD CONDENSER ON A 60 CYCLE, A-C CIRCUIT.

$$X(C) = \frac{1}{6.28 \times F \times C} = \frac{1}{6.28 \times 60 \times .000004}$$

$$= \frac{1}{0.0015072} = \underline{663 \text{ OHMS}}$$

A HENRY IS A UNIT OF INDUCTANCE, EQUAL TO THE INDUCTANCE OF A CIRCUIT IN WHICH THE VARIATION OF A CURRENT AT THE RATE OF ONE AMPERE PER SECOND INDUCES AN ELECTROMOTIVE FORCE OF ONE VOLT.

RESISTOR COLOR CODE

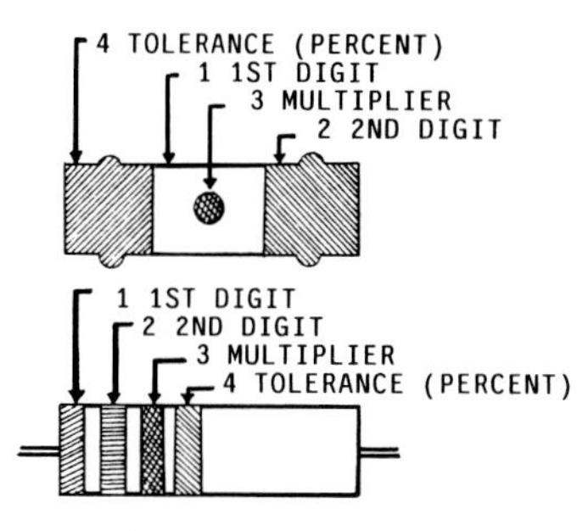

COLOR	1ST DIGIT	2ND DIGIT	MULTIPLIER	TOLERANCE (PERCENT)
BLACK	0	0	1	
BROWN	1	1	10	
RED	2	2	100	
ORANGE	3	3	1,000	
YELLOW	4	4	10,000	
GREEN	5	5	100,000	
BLUE	6	6	1,000,000	
VIOLET	7	7	10,000,000	
GRAY	8	8	100,000,000	
WHITE	9	9	1,000,000,000	
GOLD			.1	± 5 %
SILVER			.01	± 10 %
NO COLOR				± 20 %

U.S. WEIGHTS AND MEASURES

LINEAR MEASURE

			1 INCH	=	2.540 CENTIMETERS
12	INCHES	=	1 FOOT	=	3.048 DECIMETERS
3	FEET	=	1 YARD	=	9.144 DECIMETERS
5.5	YARDS	=	1 ROD, POLE, OR PERCH	=	5.029 METERS
40	RODS	=	1 FURLONG	=	2.018 HECTOMETERS
8	FURLONGS	=	1 MILE	=	1.609 KILOMETERS

MILE MEASUREMENTS

1	STATUTE MILE	=	5,280 FEET
1	SCOTS MILE	=	5,952 FEET
1	IRISH MILE	=	6,720 FEET
1	RUSSIAN VERST	=	3,504 FEET
1	ITALIAN MILE	=	4,401 FEET
1	SPANISH MILE	=	15,084 FEET

OTHER LINEAR MEASUREMENTS

1	HAND	=	4 INCHES
1	SPAN	=	9 INCHES
1	CHAIN	=	22 YARDS
1	KNOT	=	1 NAUTICAL MILE
		=	6080 FEET

1	LINK	=	7.92	INCHES
1	FATHOM	=	6	FEET
1	FURLONG	=	10	CHAINS
1	CABLE	=	608	FEET

SQUARE MEASURE

144	SQUARE INCHES	=	1 SQUARE FOOT
9	SQUARE FEET	=	1 SQUARE YARD
30-1/4	SQUARE YARDS	=	1 SQUARE ROD
		=	1 SQUARE POLE
		=	1 SQUARE PERCH
40	RODS	=	1 ROOD
4	ROODS	=	1 ACRE
640	ACRES	=	1 SQUARE MILE
1	SQUARE MILE	=	1 SECTION
36	SECTIONS	=	1 TOWNSHIP

CUBIC OR SOLID MEASURE

1	CU. FOOT	=	1728	CU. INCHES
1	CU. YARD	=	27	CU. FEET
1	CU. FOOT	=	7.48	GALLONS
1	GALLON (WATER)	=	8.34	LBS.
1	GALLON (U.S.)	=	231	CU. INCHES OF WATER
1	GALLON (IMPERIAL)	=	277-1/4	CU. INCHES OF WATER

U.S. WEIGHTS AND MEASURES

LIQUID MEASURE

1 PINT	=	4	GILLS	
1 QUART	=	2	PINTS	
1 GALLON	=	4	QUARTS	
1 FIRKIN	=	9	GALLONS	(ALE OR BEER)
1 BARREL	=	42	GALLONS	(PETROLEUM OR CRUDE OIL)

DRY MEASURE

1 QUART	=	2 PINTS
1 PECK	=	8 QUARTS
1 BUSHEL	=	4 PECKS

WEIGHT MEASUREMENT (MASS)

A. AVOIRDUPOIS WEIGHT:

1 OUNCE	=	16	DRAMS
1 POUND	=	16	OUNCES
1 HUNDREDWEIGHT	=	100	POUNDS
1 TON	=	2000	POUNDS

B. TROY WEIGHT:

1 CARAT	=	3.17	GRAINS
1 PENNYWEIGHT	=	20	GRAINS
1 OUNCE	=	20	PENNYWEIGHTS
1 POUND	=	12	OUNCES
1 LONG HUNDRED-WEIGHT	=	112	POUNDS
1 LONG TON	=	20	LONG HUNDREDWEIGHTS
	=	2240	POUNDS

C. APOTHECARIES WEIGHT:

1 SCRUPLE	=	20 GRAINS	=	1.296	GRAMS
1 DRAM	=	3 SCRUPLE	=	3.888	GRAMS
1 OUNCE	=	8 DRAMS	=	31.1035	GRAMS
1 POUND	=	12 OUNCES	=	373.2420	GRAMS

D. KITCHEN WEIGHTS AND MEASURES:

1 U.S. PINT	=	16	FL. OUNCES	
1 STANDARD CUP	=	8	FL. OUNCES	
1 TABLESPOON	=	0.5	FL. OUNCES	(15 CU. CMS.)
1 TEASPOON	=	0.16	FL. OUNCES	(5 CU. CMS.)

METRIC SYSTEM

PREFIXES:

A.	MEGA	=	1,000,000
B.	KILO	=	1,000
C.	HECTO	=	100
D.	DEKA	=	10
E.	DECI	=	0.1
F.	CENTI	=	0.01
G.	MILLI	=	0.001
H.	MICRO	=	0.000001

LINEAR MEASURE:

THE UNIT IS THE METER = 39.37 INCHES:

1 CENTIMETER	=	10 MILLIMETERS	=	0.3937011 IN.
1 DECIMETER	=	10 CENTIMETERS	=	3.9370113 INS.
1 METER	=	10 DECIMETERS	=	1.0936143 YDS.
			=	3.2808429 FT.
1 DEKAMETER	=	10 METERS	=	10.936143 YDS.
1 HECTOMETER	=	10 DEKAMETERS	=	109.36143 YDS.
1 KILOMETER	=	10 HECTOMETERS	=	0.62137 MILE
1 MYRIAMETER	=	10,000 METERS		

SQUARE MEASURE:

THE UNIT IS THE SQUARE METER = 1549.9969 SQ. INCHES:

1 SQ. CENTIMETER	=	100 SQ. MILLIMETERS	=	0.1550 SQ. IN.
1 SQ. DECIMETER	=	100 SQ. CENTIMETERS	=	15.550 SQ. INS.
1 SQ. METER	=	100 SQ. DECIMETERS	=	10.7639 SQ. FT.
1 SQ. DEKAMETER	=	100 SQ. METERS	=	119.60 SQ. YDS
1 SQ. HECTOMETER	=	100 SQ. DEKAMETERS		
1 SQ. KILOMETER	=	100 SQ. HECTOMETERS		

THE UNIT IS THE "ARE" = 100 SQ. METERS:

1 CENTIARE	=	10 MILLIARES	=	10.7643 SQ. FT.
1 DECIARE	=	10 CENTIARES	=	11.96033 SQ. YDS.
1 ARE	=	10 DECIARES	=	119.6033 SQ. YDS.
1 DEKARE	=	10 ARES	=	0.247110 ACRES
1 HEKTARE (HECTO-ARE)	=	10 DEKARES	=	2.471098 ACRES
1 SQ. KILOMETER	=	100 HEKTARES	=	0.38611 SQ. MILE

CUBIC MEASURE:

THE UNIT IS THE "STERE" = 61,025.38659 CU. INS.:

1 DECISTERE	=	10 CENTISTERES	=	3.531562 CU. FT.
1 STERE	=	10 DECISTERES	=	1.307986 CU. YDS.
1 DEKASTERE	=	10 STERES	=	13.07986 CU. YDS.

METRIC SYSTEM

CUBIC MEASURE:

THE UNIT IS THE "METER" = 39.37 INS.:

1 CU. CENTIMETER	= 1000 CU. MILLIMETERS	=	0.06125 CU. IN.
1 CU. DECIMETER	= 1000 CU. CENTIMETERS	=	61.1250 CU. INS.
1 CU. METER	= 1000 CU. DECIMETERS	=	35.3156 CU. FT.
	= 1 STERE	=	1.30797 CU. YDS.

1 CU. CENTIMETER (WATER)	=	1 GRAM
1000 CU CENTIMETERS (WATER) = 1 LITER	=	1 KILOGRAM
1 CU. METER (1000 LITERS)	=	1 METRIC TON

MEASURES OF WEIGHT:

THE UNIT IS THE GRAM = 0.035274 OUNCES:

1 MILLIGRAM			=	0.015432	GRAINS
1 CENTIGRAM	=	10 MILLIGRAMS	=	0.15432	GRAINS
1 DECIGRAM	=	10 CENTIGRAMS	=	1.5432	GRAINS
1 GRAM	=	10 DECIGRAMS	=	15.4323	GRAINS
1 DEKAGRAM	=	10 GRAMS	=	5.6438	DRAMS
1 HECTOGRAM	=	10 DEKAGRAMS	=	3.5274	OUNCES
1 KILOGRAM	=	10 HECTOGRAMS	=	2.2046223	POUNDS
1 MYRIAGRAM	=	10 KILOGRAMS	=	22.046223	POUNDS
1 QUINTAL	=	10 MYRIAGRAMS	=	1.986412	CWT.
1 METRIC TON	=	10 QUINTAL	=	2,204.622	POUNDS

1 GRAM	=	0.56438 DRAMS
1 DRAM	=	1.77186 GRAMS
	=	27.3438 GRAINS
1 METRIC TON	=	2,204.6223 POUNDS

MEASURE OF CAPACITY:

THE UNIT IS THE "LITER" = 1.0567 LIQUID QUARTS:

1 CENTILITER	=	10 MILLILITERS	=	0.338	FLUID OUNCES
1 DECILITER	=	10 CENTILITERS	=	3.38	FLUID OUNCES
1 LITER	=	10 DECILITERS	=	33.8	FLUID OUNCES
1 DEKALITER	=	10 LITERS	=	0.284	BUSHEL
1 HECTOLITER	=	10 DEKALITERS	=	2.84	BUSHELS
1 KILOLITER	=	10 HECTOLITERS	=	264.2	GALLONS

NOTE: $\frac{\text{KILOMETERS}}{8} \times 5 = \text{MILES}$ $\frac{\text{MILES}}{5} \times 8 = \text{KILOMETERS}$

TWO-WAY CONVERSION TABLE

To convert from the unit of measure in column B to the unit of measure in column C, multiply the number of units in column B by the multiplier in column D. To convert from column C to B, use the multiplier in column A.

EXAMPLE: To convert 1000 BTU's to CALORIES, find the "Btu-Calorie" combination in columns B and C. "Btu" is in column B and "Calorie" is in column C; so we are converting from B to C. Therefore, we use column D multiplier. 1000 Btu's x 251.996 = 251,996 Calories.

To convert 251,996 Calories to Btu's, use the same "Btu-Calorie" combination. But, this time you are converting from C to B. Therefore, use column A multiplier. 251,996 Calories x .0039683 = 1,000 Btu's.

To convert from C to B, Multiply by: | To convert from B to C, Multiply by:

A	B	C	D
.0295	Atmosphere	Foot of H_2O	33.89854
.0334	Atmosphere	Inch of Hg	29.92126
.0680	Atmosphere	Pound force/sq. in.	14.69595
3.96832×10^{-3}	Btu	Calorie	251.996
1.28507×10^{-3}	Btu	Foot-pound force	778.169
2544.43	Btu	Horsepower-hour	3.9301×10^{-4}
9.4781×10^{-4}	Btu	Joule	1055.056
3412.14	Btu	Kilowatt-hour	2.930×10^{-4}
3412.1425	Btu/hour	Kilowatt	2.93×10^{-4}
3.412	Btu/hour	Watt	0.293071
.23809	Btu/minute	Calorie/second	4.19993
42.4072	Btu/minute	Horsepower	.0235809
.0568	Btu/minute	Watt	17.5843
.238846	Calorie	Joule	4.1868
980.665	Dyne	Gram Force	1.0197×10^{-3}
100,000	Dyne	Newton	1×10^{-5}
1	Dyne centimeter	Erg	1
1.35558×10^{7}	Erg	Foot pound force	7.37×10^{-8}
3.6×10^{13}	Erg	Kilowatt-hour	2.777×10^{-14}
1.0×10^{7}	Erg/second	Watt	1.0×10^{-7}
.0929	Foot candle	Lux	10.76391
1.1329	Foot of H_2O	Inch of Hg	.882671
1.98×10^{6}	Foot pnd force	Horsepower-hour	5.05051×10^{-7}
.7375	Foot pnd force	Joule	1.35582

TWO-WAY CONVERSION TABLE

To convert from C to B, Multiply by: (A) — To convert from B to C, Multiply by: (D)

A	B	C	D
$2.65522x10^6$	Foot pnd force	Kilowatt-hour	$3.76616x10^{-7}$
2655.22	Foot pnd force	Watt-hour	$3.76616x10^{-4}$
$2.6552x10^6$	Foot pnd force/hour	Kilowatt	$3.766x10^{-7}$
33,000	Foot pnd force/minute	Horsepower	$3.0303x10^{-5}$
44253.7	Foot pnd force/minute	Kilowatt	$2.25x10^{-5}$
44.2537	Foot pnd force/minute	Watt	0.0225970
550	Foot pnd force/second	Horsepower	$1.81818x10^{-3}$
737.562	Foot pnd force/second	Kilowatt	$1.355818x10^{-3}$
1.34102	Horsepower	Kilowatt	.7457
.00134	Horsepower	Watt	745.7
453.6	Gram	Pound mass	.0022046
13.5951	Inch of H_2O	Inch of Hg	0.07355
$3.6x10^6$	Joule	Kilowatt-hour	$2.7777x10^{-7}$
3600	Joule	Watt hour	$2.7777x10^{-4}$
1	Joule	Watt second	1
4.448	Newton	Pound force	.2248
32.174	Pound	Slug	.03108

1 BTU RAISES 1 LB. OF WATER 1°C
1 GRAM CALORIE RAISES 1 GRAM OF WATER 1°C
1 CIRCULAR MIL EQUALS 0.7854 SQ. MIL
1 SQ. MIL EQUALS 1.27 CIR. MILS
1 MIL EQUALS 0.001 INS.

TO DETERMINE CIRCULAR MIL OF A CONDUCTOR:

ROUND CONDUCTOR............CM = (DIAMETER IN MILS)^2

BUS BAR...............................CM = $\frac{\text{WIDTH (MILS) X THICKNESS (MILS)}}{0.7854}$

NOTES:

1 MILLIMETER = 39.37 MILS
1 CIR. MILLIMETER = 1550 CIR. MILS
1 SQ. MILLIMETER = 1974 CIR. MILS

METALS

METAL	SYMB.	SPEC. GRAV.	MELT POINT C°	MELT POINT F°	ELEC. COND. % COPPER	LBS. CU."
ALUMINUM	AL	2.710	660	1220	64.9	.0978
ANTIMONY	SB	6.620	630	1167	4.42	.2390
ARSENIC	AS	5.730	----	----	4.9	.2070
BERYLLIUM	BE	1.830	1280	2336	9.32	.0660
BISMUTH	BI	9.800	271	520	1.50	.3540
BRASS (70-30)		8.510	900	1652	28.0	.3070
BRONZE (5% SN)		8.870	1000	1382	18.0	.3200
CADMIUM	CD	8.650	321	610	22.7	.3120
CALCIUM	CA	1.550	850	1562	50.1	.0560
COBALT	CO	8.900	1495	2723	17.8	.3210
COPPER	CU					
ROLLED		8.890	1083	1981	100.00	.3210
TUBING		8.950	----	----	100.00	.3230
GOLD	AU	19.30	1063	1945	71.2	.6970
GRAPHITE		2.25	3500	6332	10^{-3}	.0812
INDIUM	IN	7.30	156	311	20.6	.2640
IRIDIUM	IR	22.40	2450	4442	32.5	.8090
IRON	FE	7.20	1200 TO 1400	2192 TO 2552	17.6	.2600
MALLEABLE		7.20	1500 TO 1600	2732 TO 2912	10	.2600
WROUGHT		7.70	1500 TO 1600	2732 TO 2912	10	.2780
LEAD	PB	11.40	327	621	8.35	.4120
MAGNESIUM	MG	1.74	651	1204	38.7	.0628
MANGANESE	MN	7.20	1245	2273	0.9	.2600
MERCURY	HG	13.65	-38.9	-37.7	1.80	.4930
MOLYBDENUM	MO	10.20	2620	4748	36.1	.3680
MONEL (63-37)		8.87	1300	2372	3.0	.3200
NICKEL	NI	8.90	1452	2646	25.0	.3210
PHOSPHORUS	P	1.82	44.1	111.4	10^{-17}	.0657
PLATINUM	PT	21.46	1773	3221	17.5	.7750
POTASSIUM	K	0.860	62.3	144.1	28	.0310
SELENIUM	SE	4.81	220	428	14.4	.1740
SILICON	SI	2.40	1420	2588	10^{-5}	.0866
SILVER	AG	10.50	960	1760	106	.3790
STEEL (CARBON)		7.84	1330 TO 1380	2436 TO 2516	10	.2830
STAINLESS						
(18-8)		7.92	1500	2732	2.5	.2860
(13-CR)		7.78	1520	2768	3.5	.2810
(18-CR)		7.73	1500	2732	3.0	.2790
TANTALUM	TA	16.6	2900	5414	13.9	.5990

METALS

METAL	SYMB.	SPEC. GRAV.	MELT POINT		ELEC. COND. % COPPER	LBS. CU."
			C°	F°		
TELLURIUM	TE	6.2	450	846	10^{-5}	.2240
THORIUM	TH	11.70	1845	3353	9.10	.422
TIN	SN	7.30	232	449	15.00	.264
TITANIUM	TI	4.50	1800	3272	2.10	.162
TUNGSTEN	W	19.30	3410	----	31.50	.697
URANIUM	U	18.70	1130	2066	2.80	.675
VANADIUM	V	5.96	1710	3110	6.63	.215
ZINC	ZN	7.14	419	786	29.10	.258
ZIRCONIUM	ZR	6.40	1700	3092	4.20	.231

SPECIFIC RESISTANCE (K)

THE SPECIFIC RESISTANCE (K) OF A MATERIAL IS THE RESISTANCE OFFERED BY A WIRE OF THIS MATERIAL WHICH IS ONE FOOT LONG WITH A DIAMETER OF ONE MIL.

MATERIAL	"K"	MATERIAL	"K"
BRASS	43.0	ALUMINUM	17.0
CONSTANTAN	295	MONEL	253
COPPER	10.8	NICHROME	600
GERMAN SILVER 18%	200	NICKEL	947
GOLD	14.7	TANTALUM	93.3
IRON (PURE)	60.0	TIN	69.0
MAGNESIUM	276	TUNGSTEN	34.0
MANGANIN	265	SILVER	9.7

NOTE: 1. THE RESISTANCE OF A WIRE IS DIRECTLY PROPORTIONAL TO THE SPECIFIC RESISTANCE OF THE MATERIAL.

2. "K" = SPECIFIC RESISTANCE

CENTIGRADE AND FAHRENHEIT THERMOMETER SCALES

DEG-C	DEG-F	DEG-C	DEG-F	DEG-C	DEG-F	DEG-C	DEG-F
0	32						
1	33.8	26	78.8	51	123.8	76	168.8
2	35.6	27	80.6	52	125.6	77	170.6
3	37.4	28	82.4	53	127.4	78	172.4
4	39.2	29	84.2	54	129.2	79	174.2
5	41	30	86	55	131	80	176
6	42.8	31	87.8	56	132.8	81	177.8
7	44.6	32	89.6	57	134.6	82	179.6
8	46.4	33	91.4	58	136.4	83	181.4
9	48.2	34	93.2	59	138.2	84	183.2
10	50	35	95	60	140	85	185
11	51.8	36	96.8	61	141.8	86	186.8
12	53.6	37	98.6	62	143.6	87	188.6
13	55.4	38	100.4	63	145.4	88	190.4
14	57.2	39	102.2	64	147.2	89	192.2
15	59	40	104	65	149	90	194
16	60.8	41	105.8	66	150.8	91	195.8
17	62.6	42	107.6	67	152.6	92	197.6
18	64.4	43	109.4	68	154.4	93	199.4
19	66.2	44	111.2	69	156.2	94	201.2
20	68	45	113	70	158	95	203
21	69.8	46	114.8	71	159.8	96	204.8
22	71.6	47	116.6	72	161.6	97	206.6
23	73.4	48	118.4	73	163.4	98	208.4
24	75.2	49	120.2	74	165.2	99	210.2
25	77	50	122	75	167	100	212

1. TEMP. C° = 5/9 × (TEMP. F° - 32)
2. TEMP. F° = (9/5 × TEMP. C°) + 32
3. AMBIENT TEMPERATURE IS THE TEMPERATURE OF THE SURROUNDING COOLING MEDIUM
4. RATED TEMPERATURE RISE IS THE PERMISSIBLE RISE IN TEMPERATURE ABOVE AMBIENT WHEN OPERATING UNDER LOAD.

USEFUL MATH FORMULAS

RIGHT TRIANGLE

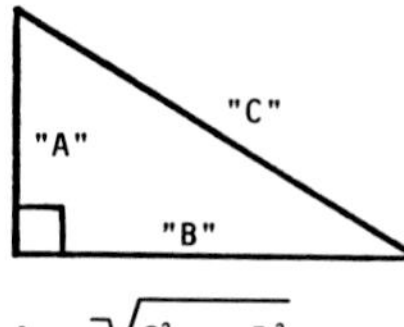

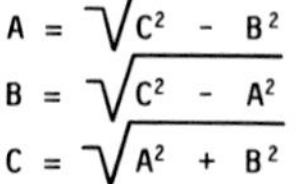

$A = \sqrt{C^2 - B^2}$

$B = \sqrt{C^2 - A^2}$

$C = \sqrt{A^2 + B^2}$

OBTUSE TRIANGLE

C-1 C-2 A B-1 B-2

SOLVE AS TWO RIGHT TRIANGLES

SPHERE

AREA = $D^2 \times 3.1416$
VOLUME = $D^3 \times 0.5236$

CYLINDRICAL

VOLUME = AREA OF END × HEIGHT

CONE

VOLUME = AREA OF END × HEIGHT / 3

"L" "W"

ELLIPTICAL

SOLVE THE SAME AS CYLINDRICAL

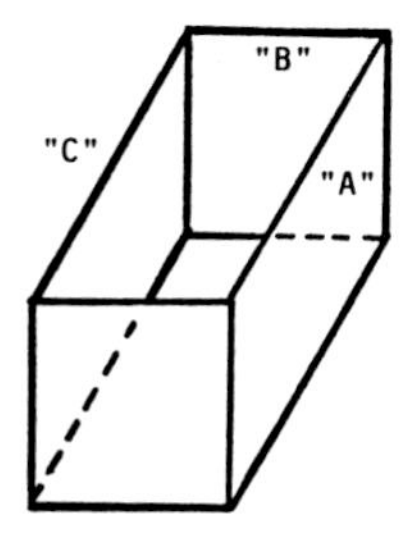

VOLUME = A × B × C

THE CIRCLE

DEFINITION: A CLOSED PLANE CURVE HAVING EVERY POINT AN EQUAL DISTANCE FROM A FIXED POINT WITHIN THE CURVE.

CIRCUMFERENCE : THE DISTANCE AROUND A CIRCLE.
DIAMETER : THE DISTANCE ACROSS A CIRCLE THROUGH THE CENTER.
RADIUS : THE DISTANCE FROM THE CENTER TO THE EDGE OF A CIRCLE.
ARC : A PART OF THE CIRCUMFERENCE.
CHORD : A STRAIGHT LINE CONNECTING THE ENDS OF AN ARC.
SEGMENT : AN AREA BOUNDED BY AN ARC AND A CHORD.
SECTOR : A PART OF CIRCLE ENCLOSED BY TWO RADII AND THE ARC WHICH THEY CUT OFF.

CIRCUMFERENCE OF A CIRCLE = 3.1416 × 2 × RADIUS
AREA OF A CIRCLE = 3.1416 × RADIUS × RADIUS
ARC LENGTH = DEGREES IN ARC × RADIUS × 0.01745
RADIUS LENGTH = ONE HALF LENGTH OF DIAMETER
SECTOR AREA = ONE HALF LENGTH OF ARC × RADIUS

CHORD LENGTH = $2\sqrt{A \times B}$
SEGMENT AREA = SECTOR AREA MINUS TRIANGLE AREA.

NOTE:

3.1416 × 2 × R = 360 DEGREES

$\frac{3.1416 \times 2 \times R}{360}$ OR

0.0087266 × 2 × R OR

0.01745 × R = 1 DEGREE

THIS GIVES US THE ARC FORMULA

DEGREES × RADIUS × 0.01745 = DEVELOPED LENGTH

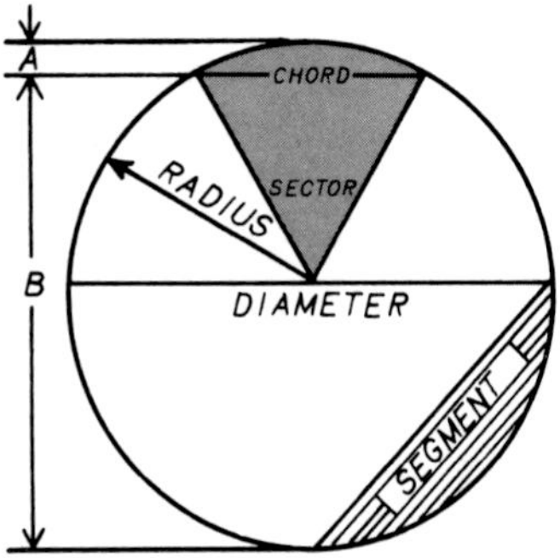

EXAMPLE:

FOR A NINETY DEGREE CONDUIT BEND, HAVING A RADIUS OF 17.25":

90 × 17.25" × 0.01745 = DEVELOPED LENGTH

27" = DEVELOPED LENGTH

FRACTIONS

DEFINITIONS:

A. A FRACTION IS A QUANTITY LESS THAN A UNIT.

B. A NUMERATOR IS THE TERM OF A FRACTION INDICATING HOW MANY OF THE PARTS OF A UNIT ARE TO BE TAKEN. IN A COMMON FRACTION IT APPEARS ABOVE OR TO THE LEFT OF THE LINE.

C. A DENOMINATOR IS THE TERM OF A FRACTION INDICATING THE NUMBER OF EQUAL PARTS INTO WHICH THE UNIT IS DIVIDED. IN A COMMON FRACTION IT APPEARS BELOW OR TO THE RIGHT OF THE LINE.

D. EXAMPLES: (1.) $\frac{1}{2} \begin{matrix}\rightarrow \\ = \\ \rightarrow\end{matrix} \frac{\text{NUMERATOR}}{\text{DENOMINATOR}} = \text{FRACTION}$

(2.) NUMERATOR → 1/2 ← DENOMINATOR

TO ADD OR SUBTRACT:

TO SOLVE 1/2 - 2/3 + 3/4 - 5/6 + 7/12 = ?

A. DETERMINE THE LOWEST COMMON DENOMINATOR THAT EACH OF THE DENOMINATORS 2, 3, 4, 6, AND 12 WILL DIVIDE INTO AN EVEN NUMBER OF TIMES.

THE LOWEST COMMON DENOMINATOR IS 12.

B. WORK ONE FRACTION AT A TIME USING THE FORMULA

$\frac{\text{COMMON DENOMINATOR}}{\text{DENOMINATOR OF FRACTION}}$ TIMES NUMERATOR OF FRACTION

(1.) 12/2 × 1 = 6 × 1 = 6 1/2 BECOMES 6/12

(2.) 12/3 × 2 = 4 × 2 = 8 2/3 BECOMES 8/12

(3.) 12/4 × 3 = 3 × 3 = 9 3/4 BECOMES 9/12

(4.) 12/6 × 5 = 2 × 5 = 10 5/6 BECOMES 10/12

(5.) 7/12 REMAINS 7/12

CONTINUED NEXT PAGE

TO ADD OR SUBTRACT (CONTINUED):

C. WE CAN NOW CONVERT THE PROBLEM FROM ITS ORIGINAL FORM TO ITS NEW FORM USING 12 AS THE COMMON DENOMINATOR.

$1/2 \; - \; 2/3 \; + \; 3/4 \; - \; 5/6 \; + \; 7/12$ = ORIGINAL FORM

$\dfrac{6 - 8 + 9 - 10 + 7}{12}$ = PRESENT FORM

$\dfrac{+22 - 18}{12} = \dfrac{+4}{12} = \dfrac{1}{3}$ REDUCED TO LOWEST FORM

D. TO CONVERT FRACTIONS TO DECIMAL FORM SIMPLY DIVIDE THE NUMERATOR OF THE FRACTION BY THE DENOMINATOR OF THE FRACTION.

EXAMPLE: 1 DIVIDED BY 3 = 0.33 = ANS.

TO MULTIPLY:

A. THE NUMERATOR OF FRACTION #1 TIMES THE NUMERATOR OF FRACTION #2 IS EQUAL TO THE NUMERATOR OF THE PRODUCT.

B. THE DENOMINATOR OF FRACTION #1 TIMES THE DENOMINATOR OF FRACTION #2 IS EQUAL TO THE DENOMINATOR OF THE PRODUCT.

C. EXAMPLE:

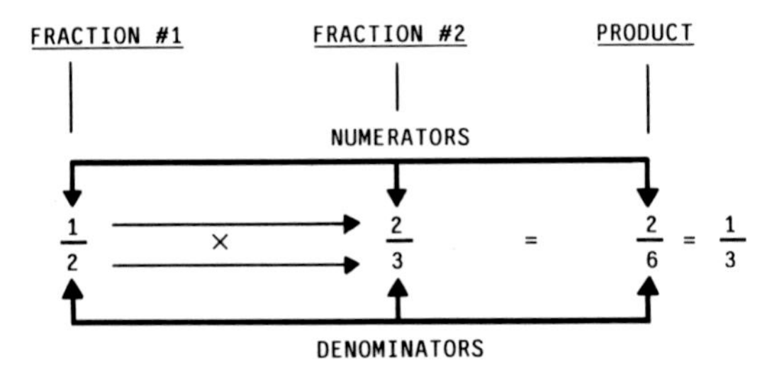

TO CHANGE 1/3 TO DECIMAL FORM, DIVIDE 1 BY 3 = 0.33

FRACTIONS

TO DIVIDE:

A. THE NUMERATOR OF FRACTION #1 TIMES THE DENOMINATOR OF FRACTION #2 IS EQUAL TO THE NUMERATOR OF THE QUOTIENT.

B. THE DENOMINATOR OF FRACTION #1 TIMES THE NUMERATOR OF FRACTION #2 IS EQUAL TO THE DENOMINATOR OF THE QUOTIENT.

C. EXAMPLE:

FRACTION #1 FRACTION #2 QUOTIENT

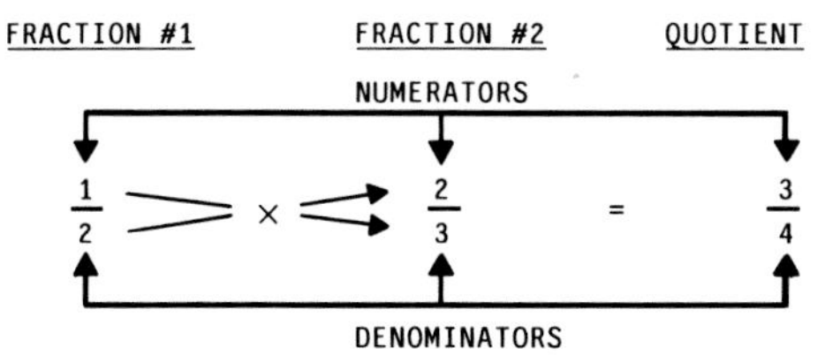

TO CHANGE 3/4 TO DECIMAL FORM, DIVIDE 3 BY 4 = 0.75

EQUATIONS

THE WORD EQUATION MEANS EQUAL OR THE SAME AS.

EXAMPLE: $2 \times 10 = 4 \times 5$
$20 = 20$

RULES:

A. THE SAME NUMBER MAY BE ADDED TO BOTH SIDES OF AN EQUATION WITHOUT CHANGING ITS VALUES.

EXAMPLE: $(2 \times 10) + 3 = (4 \times 5) + 3$
$23 = 23$

B. THE SAME NUMBER MAY BE SUBTRACTED FROM BOTH SIDES OF AN EQUATION WITHOUT CHANGING ITS VALUES.

EXAMPLE: $(2 \times 10) - 3 = (4 \times 5) - 3$
$17 = 17$

C. BOTH SIDES OF AN EQUATION MAY BE DIVIDED BY THE SAME NUMBER WITHOUT CHANGING ITS VALUES.

EXAMPLE: $$\frac{2 \times 10}{20} = \frac{4 \times 5}{20}$$

$$1 = 1$$

D. BOTH SIDES OF AN EQUATION MAY BE MULTIPLIED BY THE SAME NUMBER WITHOUT CHANGING ITS VALUES.

EXAMPLE: $3 \times (2 \times 10) = 3 \times (4 \times 5)$
$60 = 60$

E. TRANSPOSITION:

THE PROCESS OF MOVING A QUANTITY FROM ONE SIDE OF AN EQUATION TO THE OTHER SIDE OF AN EQUATION BY CHANGING ITS SIGN OF OPERATION IS TRANSPOSING.

1. A TERM MAY BE TRANSPOSED IF ITS SIGN IS CHANGED FROM PLUS (+) TO MINUS (-), OR FROM MINUS (-) TO PLUS (+).

EXAMPLES

+ TO -

$X + 5 = 25$
$X = 25 - 5$
$X = 20$

- TO +

$X - 5 = 25$
$X = 25 + 5$
$X = 30$

EQUATIONS

RULES:

E. TRANSPOSITION:

2. A MULTIPLIER MAY BE REMOVED FROM ONE SIDE OF AN EQUATION BY MAKING IT A DIVISOR IN THE OTHER, OR A DIVISOR MAY BE REMOVED FROM ONE SIDE OF AN EQUATION BY MAKING IT A MULTIPLIER IN THE OTHER.

EXAMPLE: MULTIPLIER FROM ONE SIDE OF EQUATION BECOMES DIVISOR IN OTHER SIDE OF THE EQUATION.

EXAMPLE: $4X = 40$ BECOMES $X = \frac{40}{4}$

DIVISOR FROM ONE SIDE OF EQUATION BECOMES MULTIPLIER IN OTHER SIDE OF THE EQUATION.

EXAMPLE: $\frac{X}{4} = 10$ BECOMES $X = 4 \times 10$

SIGNS:

A. ADDITION:

1. RULE: USE THE SIGN OF THE LARGER AND ADD.

EXAMPLES:

+ 3	- 2	+ 3	- 3
- 2	+ 3	+ 2	- 2
+ 1	+ 1	+ 5	- 5

B. SUBTRACTION:

1. RULE: CHANGE THE SIGN OF THE SUBTRAHEND AND PROCEED AS IN ADDITION:

EXAMPLES:

+ 3	- 2	+ 3	- 3
- 2	+ 3	+ 2	- 2

CHANGE SUBTRAHEND AND ADD

+ 3	- 2	+ 3	- 3
+ 2	- 3	- 2	+ 2
+ 5	- 5	+ 1	- 1

EQUATIONS

3. SIGNS (CONTINUED):

C. MULTIPLICATION:

1. THE PRODUCT OF ANY TWO NUMBERS HAVING LIKE SIGNS IS POSITIVE. THE PRODUCT OF ANY TWO NUMBERS HAVING UNLIKE SIGNS IS NEGATIVE.

EXAMPLE: $(+3) \times (-2) = -6$ $\quad (-3) \times (+2) = -6$

$(+3) \times (+2) = +6$ $\quad (-3) \times (-2) = +6$

D. DIVISION:

1. IF THE DIVISOR AND DIVIDEND HAVE LIKE SIGNS, THE SIGN OF THE QUOTIENT IS POSITIVE. IF THE DIVISOR AND DIVIDEND HAVE UNLIKE SIGNS, THE SIGN OF THE QUOTIENT IS NEGATIVE.

EXAMPLE:

$$\frac{+6}{-2} = -3 \qquad \frac{+6}{+2} = +3 \qquad \frac{-6}{+2} = -3 \qquad \frac{-6}{-2} = +3$$

SQUARE ROOT

1. GROUPING THE DIGITS IN A NUMBER IS ESSENTIAL IN SOLVING SQUARE ROOT PROBLEMS. START AT THE DECIMAL POINT, AND GROUP TWO TO A GROUP TO THE LEFT. IF THERE IS A DIGIT LEFT OVER AT THE EXTREME LEFT THAT DIGIT WILL BE CONSIDERED TO BE A GROUP. START AGAIN AT THE DECIMAL POINT AND GROUP TWO TO A GROUP TO THE RIGHT. IF THERE IS A DIGIT LEFT OVER AT THE EXTREME RIGHT, SIMPLY ADD A "0", WHICH WILL NOT CHANGE THE VALUE OF THE NUMBER.

   EXAMPLES: 234.567 GROUP AS 2 34 . 56 70
   2345.67 GROUP AS 23 45 . 67

2. THE CONSTANT IN SOLVING ALL SQUARE ROOT PROBLEMS WILL BE NUMBER "20".

3. SOLUTION:

```
                              1  5 .  3  1
                          _____________________
                         √ 2 34  .  56  70
20 × 1    =  20         - 1
          +   5
          =  25    |      1 34
20 × 15   = 300         - 1 25
          +   3
          = 303    |         9     56
20 × 153  = 3060          - 9     09
          +    1
          = 3061   |               47  70
                                 - 30  61
```

(STEP 1.)

THE LARGEST NUMBER THAT WILL SQUARE INTO THE FIRST GROUP IS "1".

SUBTRACT "1" FROM "2" AND BRING DOWN SECOND GROUP "34"

CONSTANT 20 × ANSWER 1 = 20
DIVIDE 134 BY 20 = 6
ADD 20 AND 6 = 26

NOTE: 26 WILL NOT DIVIDE INTO 134 SIX TIMES SO THEREFORE 25 MUST BE USED. 25 WILL DIVIDE INTO 134 FIVE TIMES. NOW OUR ANSWER IS 5 AND WE ARE READY FOR STEP 2.

(STEP 2.)

MULTIPLY 5 × 25 TO = 125
SUBTRACT 125 FROM 134 TO = 9
BRING DOWN THIRD GROUP 56 AND OUR REMAINDER = 956

CONSTANT 20 × TOTAL ANSWER 15 IS 300.

DIVIDE 956 BY 300 = 3
ADD 300 AND 3 = 303
303 WILL DIVIDE INTO 956 THREE TIMES, SO 3 BECOMES THE THIRD DIGIT IN OUR ANSWER

(STEP 3.)

3 × 303 = 909
SUBTRACT 909 FROM 956 TO EQUAL 47. BRING DOWN LAST GROUP 70 TO ESTABLISH A REMAINDER OF 4770.

(STEP 4.)

CONSTANT 20 × TOTAL ANSWER 153 = 3060.
DIVIDE 4770 BY 3060 = 1
ADD 3060 AND 1 = 3061
3061 WILL DIVIDE INTO 4770 ONE TIME, 1 BECOMES THE FOURTH DIGIT IN OUR ANSWER.

NOTE: MULTIPLY 15.31 × 15.31 = 234.396. ADD 234.396 AND 0.1709 (THE REMAINDER OF THE PROBLEM) TO EQUAL 234.567. THIS ENABLES YOU TO CHECK FOR ACCURACY.

NATURAL TRIGONOMETRIC FUNCTIONS

ANGLE	SINE	COSINE	TANGENT	COTAN.	SECANT	COSECANT	ANGLE
0	.0000	1.0000	.0000		1.0000		90
1	.0175	.9998	.0175	57.2900	1.0002	57.2987	89
2	.0349	.9994	.0349	28.6363	1.0006	28.6537	88
3	.0523	.9986	.0524	19.0811	1.0014	19.1073	87
4	.0698	.9976	.0699	14.3007	1.0024	14.3356	86
5	.0872	.9962	.0875	11.4301	1.0038	11.4737	85
6	.1045	.9945	.1051	9.5144	1.0055	9.5668	84
7	.1219	.9925	.1228	8.1443	1.0075	8.2055	83
8	.1392	.9903	.1405	7.1154	1.0098	7.1853	82
9	.1564	.9877	.1584	6.3138	1.0125	6.3925	81
10	.1736	.9848	.1763	5.6713	1.0154	5.7588	80
11	.1908	.9816	.1944	5.1446	1.0187	5.2408	79
12	.2079	.9781	.2126	4.7046	1.0223	4.8097	78
13	.2250	.9744	.2309	4.3315	1.0263	4.4454	77
14	.2419	.9703	.2493	4.0108	1.0306	4.1336	76
15	.2588	.9659	.2679	3.7321	1.0353	3.8637	75
16	.2756	.9613	.2867	3.4874	1.0403	3.6280	74
17	.2924	.9563	.3057	3.2709	1.0457	3.4203	73
18	.3090	.9511	.3249	3.0777	1.0515	3.2361	72
19	.3256	.9455	.3443	2.9042	1.0576	3.0716	71
20	.3420	.9397	.3640	2.7475	1.0642	2.9238	70
21	.3584	.9336	.3839	2.6051	1.0711	2.7904	69
22	.3746	.9272	.4040	2.4751	1.0785	2.6695	68
23	.3907	.9205	.4245	2.3559	1.0864	2.5593	67
24	.4067	.9135	.4452	2.2460	1.0946	2.4586	66
25	.4226	.9063	.4663	2.1445	1.1034	2.3662	65
26	.4384	.8988	.4877	2.0503	1.1126	2.2812	64
27	.4540	.8910	.5095	1.9626	1.1223	2.2027	63
28	.4695	.8829	.5317	1.8807	1.1326	2.1301	62
29	.4848	.8746	.5543	1.8040	1.1434	2.0627	61
30	.5000	.8660	.5774	1.7321	1.1547	2.0000	60
31	.5150	.8572	.6009	1.6643	1.1666	1.9416	59
32	.5299	.8480	.6249	1.6003	1.1792	1.8871	58
33	.5446	.8387	.6494	1.5399	1.1924	1.8361	57
34	.5592	.8290	.6745	1.4826	1.2062	1.7883	56
35	.5736	.8192	.7002	1.4281	1.2208	1.7434	55
36	.5878	.8090	.7265	1.3764	1.2361	1.7013	54
37	.6018	.7986	.7536	1.3270	1.2521	1.6616	53
38	.6157	.7880	.7813	1.2799	1.2690	1.6243	52
39	.6293	.7771	.8098	1.2349	1.2868	1.5890	51
40	.6428	.7660	.8391	1.1918	1.3054	1.5557	50
41	.6561	.7547	.8693	1.1504	1.3250	1.5243	49
42	.6691	.7431	.9004	1.1106	1.3456	1.4945	48
43	.6820	.7314	.9325	1.0724	1.3673	1.4663	47
44	.6947	.7193	.9657	1.0355	1.3902	1.4396	46
45	.7071	.7071	1.0000	1.0000	1.4142	1.4142	45
ANGLE	COSINE	SINE	COTAN.	TANGENT	COSECANT	SECANT	ANGLE

TRIGONOMETRY

TRIGONOMETRY IS THE MATHEMATICS DEALING WITH THE RELATIONS OF SIDES AND ANGLES OF TRIANGLES.

A TRIANGLE IS A FIGURE ENCLOSED BY THREE STRAIGHT SIDES. THE SUM OF THE THREE ANGLES IS 180 DEGREES. ALL TRIANGLES HAVE SIX PARTS: THREE ANGLES, AND THREE SIDES OPPOSITE THE ANGLES.

RIGHT TRIANGLES ARE TRIANGLES THAT HAVE ONE ANGLE OF NINETY DEGREES AND TWO ANGLES OF LESS THAN NINETY DEGREES.

TO HELP YOU REMEMBER THE SIX TRIGONOMETRIC FUNCTIONS; MEMORIZE

"OH HELL ANOTHER HOUR OF ANDY"

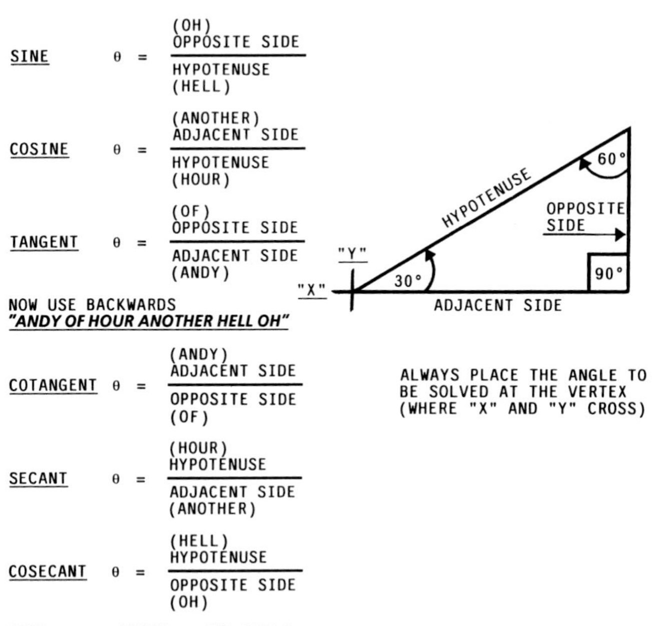

SINE $\theta = \dfrac{\text{(OH) OPPOSITE SIDE}}{\text{HYPOTENUSE (HELL)}}$

COSINE $\theta = \dfrac{\text{(ANOTHER) ADJACENT SIDE}}{\text{HYPOTENUSE (HOUR)}}$

TANGENT $\theta = \dfrac{\text{(OF) OPPOSITE SIDE}}{\text{ADJACENT SIDE (ANDY)}}$

NOW USE BACKWARDS
"ANDY OF HOUR ANOTHER HELL OH"

COTANGENT $\theta = \dfrac{\text{(ANDY) ADJACENT SIDE}}{\text{OPPOSITE SIDE (OF)}}$

SECANT $\theta = \dfrac{\text{(HOUR) HYPOTENUSE}}{\text{ADJACENT SIDE (ANOTHER)}}$

COSECANT $\theta = \dfrac{\text{(HELL) HYPOTENUSE}}{\text{OPPOSITE SIDE (OH)}}$

ALWAYS PLACE THE ANGLE TO BE SOLVED AT THE VERTEX (WHERE "X" AND "Y" CROSS)

NOTE: θ = THETA = ANY ANGLE

BENDING OFF-SETS WITH TRIGONOMETRY

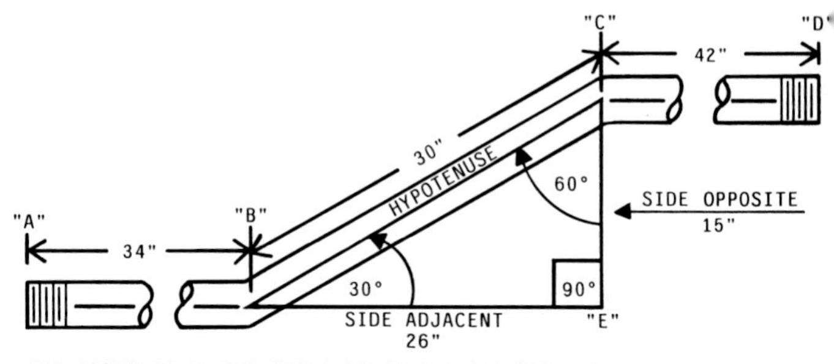

THE COSECANT OF THE ANGLE TIMES THE OFF-SET DESIRED IS EQUAL TO THE DISTANCE BETWEEN THE CENTERS OF THE BENDS.

EXAMPLE: TO MAKE A FIFTEEN INCH (15") OFF-SET: USING THIRTY (30) DEGREE BENDS:

1. USE TRIG. TABLE (PAGE 54) TO FIND THE COSECANT OF A THIRTY (30) DGREE ANGLE. WE FIND IT TO BE TWO (2).
2. MULTIPLY TWO (2) TIMES THE OFF-SET DESIRED, WHICH IS FIFTEEN (15) INCHES TO DETERMINE THE DISTANCE BETWEEN BEND "B" AND BEND "C". THE ANSWER IS THIRTY (30) INCHES.

TO MARK THE CONDUIT FOR BENDING:

1. MEASURE FROM END OF CONDUIT "A" THIRTY-FOUR (34) INCHES TO CENTER OF FIRST BEND "B", AND MARK.
2. MEASURE FROM MARK "B" THIRTY (30) INCHES TO CENTER OF SECOND BEND "C", AND MARK.
3. MEASURE FROM MARK "C" FORTY-TWO (42) INCHES TO "D", AND MARK. CUT, REAM, AND THREAD CONDUIT BEFORE BENDING.

ROLLING OFF-SETS: TO DETERMINE HOW MUCH OFF-SET IS NEEDED TO MAKE A ROLLING OFF-SET:

1. MEASURE VERTICAL REQUIRED. USE WORK TABLE (ANY SQUARE WILL DO) AND MEASURE FROM CORNER THIS AMOUNT AND MARK.
2. MEASURE HORIZONTAL REQUIRED. MEASURE NINETY DEGREES FROM THE VERTICAL LINE MEASUREMENT (STARTING IN SAME CORNER) AND MARK.
3. THE DIAGONAL DISTANCE BETWEEN THESE MARKS WILL BE THE AMOUNT OF OFF-SET REQUIRED.

NOTE: SHRINK IS HYPOTENUSE MINUS THE SIDE ADJACENT.

CHICAGO-TYPE BENDERS

NINETY DEGREE BENDING

"A" TO "C" = STUB-UP
"C" TO "D" = TAIL

"C" = BACK OF STUB-UP
"C" = BOTTOM OF CONDUIT

NOTE

THERE ARE MANY VARIATIONS OF THIS TYPE BENDER, BUT MOST MANUFACTURERS OFFER TWO SIZES.

THE SMALL SIZE SHOE TAKES 1/2", 3/4", AND 1" CONDUIT.

THE LARGE SIZE SHOE TAKES 1-1/4" AND 1-1/2" CONDUIT.

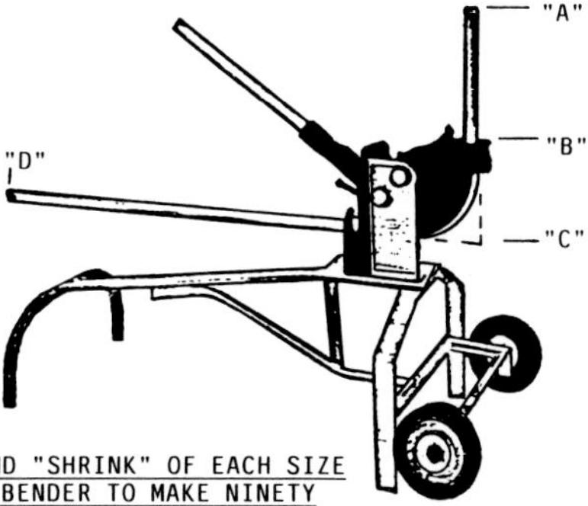

"A"
"B"
"D"
"C"

TO DETERMINE "TAKE-UP" AND "SHRINK" OF EACH SIZE CONDUIT FOR A PARTICULAR BENDER TO MAKE NINETY DEGREE BENDS.

1. USE A STRAIGHT PIECE OF SCRAP CONDUIT.

2. MEASURE EXACT LENGTH OF SCRAP CONDUIT, "A" TO "D".

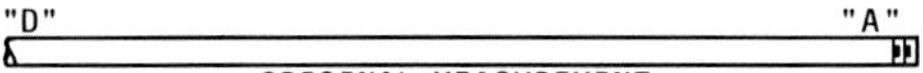

"D" "A"
ORIGINAL MEASUREMENT

3. PLACE CONDUIT IN BENDER. MARK AT EDGE OF SHOE, "B".

4. LEVEL CONDUIT, BEND NINETY, AND COUNT NUMBER OF PUMPS. MAKE AND KEEP NOTES ON EACH SIZE CONDUIT USED.

5. AFTER BENDING NINETY:

    A. DISTANCE BETWEEN "B" AND "C" IS THE TAKE-UP.

    B. ORIGINAL MEASUREMENT OF THE SCRAP PIECE OF CONDUIT SUBTRACTED FROM (DISTANCE "A" TO "C" PLUS DISTANCE "C" TO "D") IS THE SHRINK.

NOTE: BOTH TIME AND ENERGY WILL BE SAVED IF CONDUIT CAN BE CUT, REAMED, AND THREADED BEFORE BENDING.

THE SAME METHOD CAN BE USED ON HYDRAULIC BENDERS.

CHICAGO-TYPE BENDER

OFF-SETS

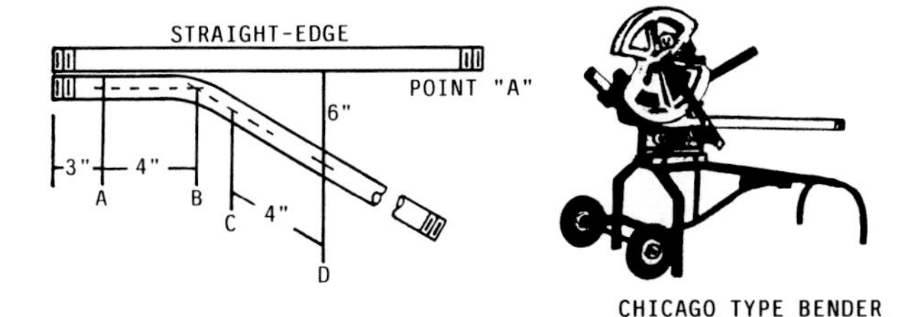

CHICAGO TYPE BENDER

EXAMPLE: TO BEND A 6" OFF-SET:

1. MAKE A MARK 3" FROM CONDUIT END. PLACE CONDUIT IN BENDER WITH MARK AT OUTSIDE EDGE OF JAW.

2. MAKE THREE FULL PUMPS, MAKING SURE HANDLE GOES ALL THE WAY DOWN TO THE STOP.

3. REMOVE CONDUIT FROM BENDER AND PLACE ALONG SIDE STRAIGHT-EDGE.

4. MEASURE 6" FROM STRAIGHT-EDGE TO CENTER OF CONDUIT. MARK POINT "D". USE SQUARE FOR ACCURACY.

5. MARK CENTER OF CONDUIT FROM BOTH DIRECTIONS THROUGH BEND AS ILLUSTRATED BY BROKEN LINE. WHERE LINES CROSS IS POINT "B".

6. MEASURE FROM "A" TO "B" TO DETERMINE DISTANCE FROM "D" TO "C". MARK "C" AND PLACE CONDUIT IN BENDER WITH MARK AT OUTSIDE EDGE OF JAW, AND WITH THE KICK POINTING DOWN. USE LEVEL TO PREVENT DOGGING CONDUIT.

7. REPEAT STEP "2".

NOTE: 1. THERE ARE SEVERAL METHODS OF BENDING RIGID CONDUIT WITH A CHICAGO TYPE BENDER, AND ANY METHOD THAT GETS THE JOB DONE IN A MINIMUM AMOUNT OF TIME WITH CRAFTSMANSHIP IS GOOD.

2. WHATEVER METHOD USED, QUALITY WILL IMPROVE WITH EXPERIENCE.

MULTI-SHOT NINETY DEGREE CONDUIT BENDING

PROBLEM:

A. TO MEASURE, THREAD, CUT, AND REAM CONDUIT BEFORE BENDING.
B. TO ACCURATELY BEND CONDUIT TO THE DESIRED HEIGHT OF THE STUB-UP (H), AND TO THE DESIRED LENGTH OF THE TAIL (L).

GIVEN:

A. SIZE OF CONDUIT, 2"
B. SPACE BETWEEN CONDUIT (CENTER TO CENTER), 6"
C. HEIGHT OF STUB-UP, 36"
D. LENGTH OF TAIL, 48"

SOLUTION:

A. TO DETERMINE RADIUS: (R)

CONDUIT # 1 (INSIDE CONDUIT) WILL USE THE MINIMUM RADIUS UNLESS OTHERWISE SPECIFIED. THE MINIMUM RADIUS IS EIGHT TIMES THE SIZE OF THE CONDUIT, PLUS ONE-HALF THE OUTSIDE DIAMETER OF THE CONDUIT. (SEE PAGE 59)

RADIUS OF CONDUIT # 1 = 8 × 2" + 1.25" = 17.25"

RADIUS OF CONDUIT # 2 = RADIUS # 1 + 6" = 23.25"

RADIUS OF CONDUIT # 3 = RADIUS # 2 + 6" = 29.25"

B. TO DETERMINE DEVELOPED LENGTH: (DL) RADIUS × 1.57 = DL

DL OF CONDUIT # 1 = R × 1.57 = 17.25" × 1.57 = 27"

DL OF CONDUIT # 2 = R × 1.57 = 23.25" × 1.57 = 36.5"

DL OF CONDUIT # 3 = R × 1.57 = 29.25" × 1.57 = 46"

C. TO DETERMINE LENGTH OF NIPPLE: (SEE PAGE 61)

LENGTH OF NIPPLE, CONDUIT # 1 = L + H + DL - 2R
= 48" + 36" + 27" - 34.5"
= 76.5" = 6' - 4.5"

LENGTH OF NIPPLE, CONDUIT # 2 = L + H + DL - 2R
= 54" + 42" + 36.5" - 46.5"
= 86" = 7' - 2"

LENGTH OF NIPPLE, CONDUIT # 3 = L + H + DL - 2R
= 60" + 48" + 46" - 58.5"
= 95.5" = 7' - 11.5"

NOTE: 1. FOR 90 DEGREE BENDS, SHRINK = 2R - DL

2. FOR OFF-SET BENDS, SHRINK = HYPOTENUSE - SIDE ADJACENT.

MULTI-SHOT NINETY DEGREE CONDUIT BENDING

LAYOUT AND BENDING:

A. TO LOCATE POINT "B", MEASURE FROM POINT "A", THE LENGTH OF THE STUB-UP MINUS THE RADIUS. ON ALL THREE CONDUIT, POINT "B" WILL BE 18.75" FROM POINT "A". (PAGE 59)

B. TO LOCATE POINT "C", MEASURE FORM POINT "D", THE LENGTH MINUS THE RADIUS, (REFER PAGE 61). ON ALL THREE CONDUIT, POINT "C" WILL BE 30.75" FROM POINT "D". (PAGE 59)

C. DIVIDE THE DEVELOPED LENGTH (POINT "B" TO POINT "C") INTO EQUAL SPACES. SPACES SHOULD NOT BE MORE THAN 1.75" TO PREVENT WRINKLING OF THE CONDUIT. ON CONDUIT # 1, SEVENTEEN SPACES OF 1.5882" EACH, WOULD GIVE US EIGHTEEN SHOTS OF 5 DEGREES EACH. REMEMBER THERE IS ALWAYS ONE LESS SPACE THAN SHOT. WHEN DETERMINING THE NUMBER OF SHOTS, CHOOSE A NUMBER THAT WILL DIVIDE INTO NINETY AN EVEN NUMBER OF TIMES.

D. IF AN ELASTIC NUMBERED TAPE IS NOT AVAILABLE TRY THE METHOD ILLUSTRATED.

A TO B = CONDUIT # 1.
DEVELOPED LENGTH = 27"

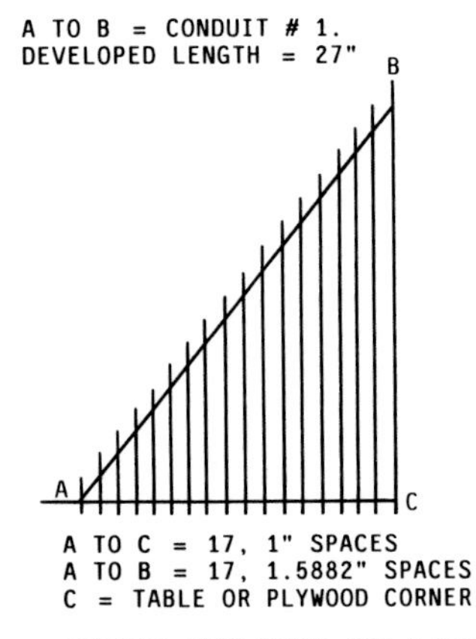

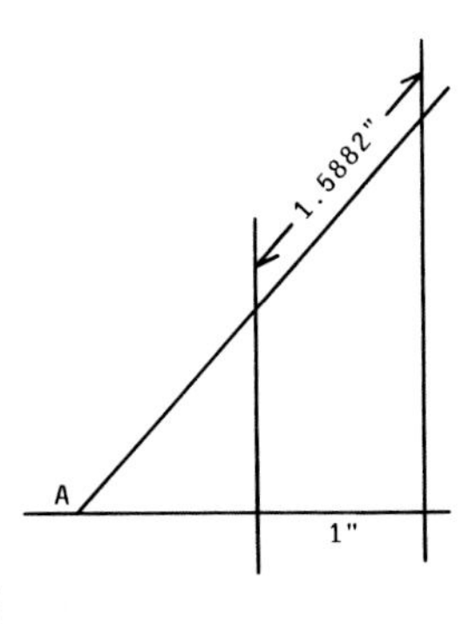

A TO C = 17, 1" SPACES
A TO B = 17, 1.5882" SPACES
C = TABLE OR PLYWOOD CORNER

MEASURE FROM POINT "C" (TABLE CORNER) 17 INCHES ALONG TABLE EDGE TO POINT "A" AND MARK. PLACE END OF RULE AT POINT "A". POINT "B" WILL BE LOCATED WHERE 27" MARK MEETS TABLE EDGE B-C. MARK ON BOARD THEN TRANSFER TO CONDUIT.

MULTI-SHOT NINETY DEGREE CONDUIT BENDING

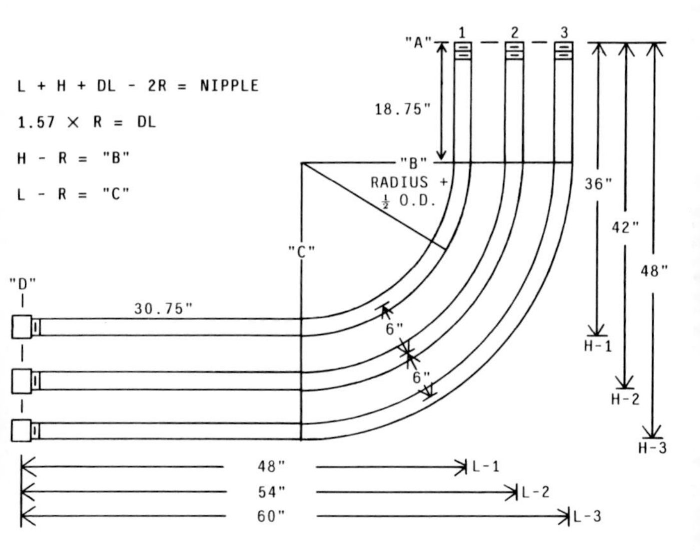

TO LOCATE POINT "B"

H #1 - RADIUS #1 = "B"
36" - 17.25" = "B"
18.75" = "B"

H #2 - RADIUS #2 = "B"
42" - 23.25" = "B"
18.75" = "B"

H #3 - RADIUS #3 = "B"
48" - 29.25" = "B"
18.75" = "B"

TO LOCATE POINT "C"

L #1 - RADIUS #1 = "C"
48" - 17.25" = "C"
30.75" = "C"

L #2 - RADIUS #2 = "C"
54" - 23.25" = "C"
30.75" = "C"

L #3 - RADIUS #3 = "C"
60" - 29.25" = "C"
18.75" = "C"

POINTS "B" AND "C" ARE THE SAME DISTANCE FROM THE END ON ALL THREE CONDUITS.

THREAD DIMENSIONS AND TAP DRILL SIZES

COARSE THREAD SERIES				FINE THREAD SERIES			
NOMINAL SIZE	THREADS PER. IN.	TAP DRILL	CLEARANCE DRILL	NOMINAL SIZE	THREADS PER. IN.	TAP DRILL	CLEARANCE DRILL
5/64"	48	47	36	0	80	3/64"	51
1/8"	40	38	29	1	72	53	47
6	32	36	25	2	64	50	42
8	32	29	16	3	56	45	36
10	24	25	13/64"	4	48	42	31
12	24	16	7/32"	1/8"	44	37	29
1/4"	20	7	17/64"	6	40	33	25
5/16"	18	F	21/64"	8	36	29	16
3/8"	16	5/16"	25/64"	10	32	21	13/64"
7/16"	14	U	29/64"	12	28	14	7/32"
1/2"	13	27/64"	33/64"	1/4"	28	3	17/64"
9/16"	12	31/64"	37/64"	5/16"	24	1	21/64"
5/8"	11	17/32"	41/64"	3/8"	24	Q	25/64"
3/4"	10	21/32"	49/64"	7/16	20	25/64	29/64"
7/8"	9	49/64"	57/64"	1/2"	20	29/64"	33/64"
1"	8	7/8"	1-1/64"	9/16"	18	33/64"	37/64"
1-1/4"	7	1-11/64"	1-17/64	5/8"	18	37/64"	41/64"
1-3/8"	6	1-19/64	1-25/64	3/4"	16	11/16"	49/64"
1-1/2"	6	1-27/64"	1-33/64"	7/8"	14	13/16"	57/64"
2"	4-1/2"	1-25/32"	2-1/32"	1	14	15/16"	1-1/64"

HOLE SAW CHART

TRADE SIZE	RIGID CONDUIT	E.M.T. CONDUIT	GREEN-FIELD	L. T. FLEX.	TRADE SIZE	RIGID CONDUIT	E.M.T. CONDUIT	GREEN-FIELD
1/2"	7/8"	3/4"	1"	1-1/8"	2-1/2"	3"	2-7/8"	2-7/8"
3/4"	1-1/8"	1"	1-1/8"	1-1/4"	3"	3-5/8"	3-1/2"	3-5/8"
1"	1-3/8"	1-1/4"	1-1/2"	1-1/2"	3-1/2"	4-1/8"	4"	4-1/8"
1-1/4"	1-3/4"	1-5/8"	1-3/4"	1-7/8"	4"	4-5/8"	4-1/2"	4-5/8"
1-1/2"	2"	1-7/8"	2"	2-1/8"	5"	5-3/4"		
2"	2-1/2	2-1/8"	2-1/2"	2-3/4"	6"	6-3/4"		

NOTE: FOR OIL TYPE PUSH BUTTON STATION USE SIZE 1-7/32" KNOCK-OUT PUNCH.

RUNNING OVER-LOAD UNITS

KIND OF MOTOR	SUPPLY SYSTEM	NUMBER AND LOCATION OF OVER-LOAD UNITS, SUCH AS TRIP COILS OR RELAYS
1- Phase ac or dc	2-wire, 1-phase ac or dc ungrounded	1 in either conductor
1- Phase ac or dc	2-wire, 1-phase ac or dc, one conductor grounded	1 in ungrounded conductor
1- Phase ac or dc	3-wire, 1-phase ac or dc, grounded neutral	1 in either ungrounded conductor
1- Phase ac	Any 3-phase	1 in ungrounded conductor
2- Phase ac	3-wire, 2-phase ac, ungrounded	2, one in each phase
2- Phase ac	3-wire, 2-phase ac, one conductor grounded	2 in ungrounded conductors
2-Phase ac	4-wire, 2-phase ac, grounded or ungrounded	2, one per phase in un-grounded conductors
2-Phase ac	5-wire, 2-phase ac, grounded neutral or ungrounded	2, one per phase in any ungrounded phase wire
3-Phase ac	Any 3-phase	3, one in each phase*

*Exception: Where protected by other approved means.

MOTOR BRANCH-CIRCUIT PROTECTIVE DEVICES

MAXIMUM RATING OR SETTING

TYPE OF MOTOR	PERCENT OF FULL-LOAD CURRENT			
	NONTIME DELAY FUSE	DUAL-ELEMENT TIME-DELAY FUSE	INSTANTANEOUS TRIP BREAKER	INVERSE TIME BREAKER
SINGLE-PHASE, ALL TYPES (NO CODE LETTER)	300	175	700	250
ALL AC SINGLE-PHASE AND POLYPHASE SQUIRREL-CAGE AND SYNCHRONOUS MOTORS WITH FULL-VOLTAGE, RESISTOR OR REACTOR STARTING				
(NO CODE LETTER)	300	175	700	250
(CODE LETTER F TO V)	300	175	700	250
(CODE LETTER B TO E)	250	175	700	200
(CODE LETTER A)	150	150	700	150
ALL AC SQUIRREL-CAGE AND SYNCHRONOUS MOTORS WITH AUTOTRANSFORMER STARTING, NOT MORE THAN 30 AMPS				
(NO CODE LETTER)	250	175	700	200
MORE THAN 30 AMPS				
(NO CODE LETTER)	200	175	700	200
(CODE LETTER F TO V)	250	175	700	200
(CODE LETTER B TO E)	200	175	700	200
(CODE LETTER A)	150	150	700	150
HIGH-REACTANCE SQUIRREL-CAGE, NOT MORE THAN 30 AMPS,				
(NO CODE LETTER)	250	175	700	250
MORE THAN 30 AMPS,				
(NO CODE LETTER)	200	175	700	200
WOUND ROTOR				
(NO CODE LETTER)	150	150	700	150
DC (CONSTANT VOLTAGE) NO MORE THAN 50 HP				
(NO CODE LETTER)	150	150	250	150
MORE THAN 50 HP				
(NO CODE LETTER)	150	150	175	150

SYNCHRONOUS MOTORS OF THE LOW-TORQUE, LOW-SPEED TYPE (USUALLY 450 RPM OR LOWER), THAT START UNLOADED, DO NOT REQUIRE A FUSE RATING OR CIRCUIT-BREAKER SETTING IN EXCESS OF 200 PERCENT OF FULL-LOAD CURRENT.

FULL-LOAD CURRENT IN AMPERES

DIRECT-CURRENT MOTORS

HP	90V	120V	180V	240V	500V	550V
1/4	4.0	3.1	2.0	1.6	—	—
1/3	5.2	4.1	2.6	2.0	—	—
1/2	6.8	5.4	3.4	2.7	—	—
3/4	9.6	7.6	4.8	3.8	—	—
1	12.2	9.5	6.1	4.7	—	—
1-1/2	—	13.2	8.3	6.6	—	—
2	—	17	10.8	8.5	—	—
3	—	25	16	12.2	—	—
5	—	40	27	20	—	—
7-1/2	—	58	—	29	13.6	12.2
10	—	76	—	38	18	16
15	—	—	—	55	27	24
20	—	—	—	72	34	31
25	—	—	—	89	43	38
30	—	—	—	106	51	46
40	—	—	—	140	67	61
50	—	—	—	173	83	75
60	—	—	—	206	99	90
75	—	—	—	255	123	111
100	—	—	—	341	164	148
125	—	—	—	425	205	185
150	—	—	—	506	246	222
200	—	—	—	675	330	294

DIRECT CURRENT MOTORS

TERMINAL MARKINGS:

TERMINAL MARKINGS ARE USED TO TAG TERMINALS TO WHICH CONNECTIONS ARE TO BE MADE FROM OUTSIDE CIRCUITS.

FACING THE END OPPOSITE THE DRIVE (COMMUTATOR END) THE STANDARD DIRECTION OF SHAFT ROTATION IS COUNTER CLOCKWISE.

A-1 AND A-2 INDICATE ARMATURE LEADS.
S-1 AND S-2 INDICATE SERIES-FIELD LEADS.
F-1 AND F-2 INDICATE SHUNT-FIELD LEADS.

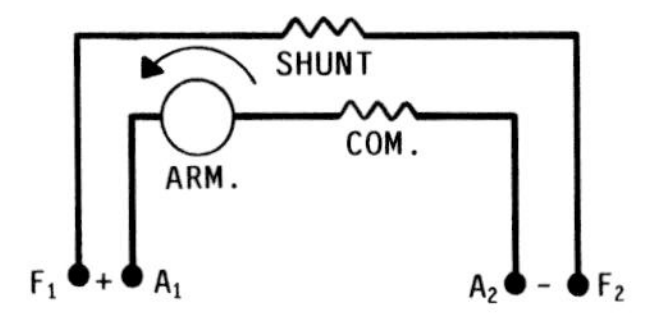

SHUNT WOUND MOTORS

TO CHANGE ROTATION, REVERSE EITHER ARMATURE LEADS OR SHUNT LEADS. DO NOT REVERSE BOTH ARMATURE AND SHUNT LEADS.

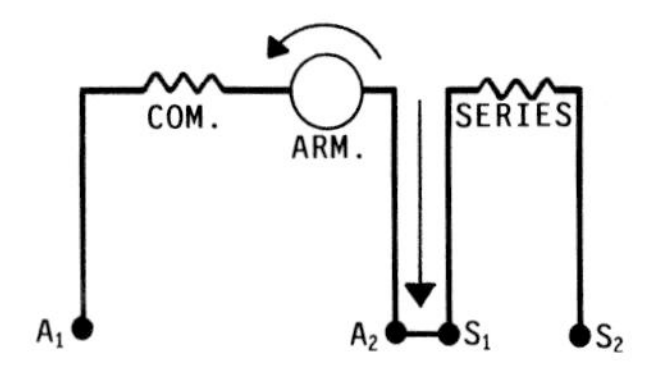

SERIES WOUND MOTORS

TO CHANGE ROTATION, REVERSE EITHER ARMATURE LEADS OR SERIES LEADS. DO NOT REVERSE BOTH ARMATURE AND SERIES LEADS.

SHUNT
SERIES
COM.
ARM.
F_1 A_1 A_2 S_1 S_2 F_2

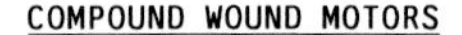

COMPOUND WOUND MOTORS

TO CHANGE ROTATION, REVERSE EITHER ARMATURE LEADS OR BOTH THE SERIES AND SHUNT LEADS. DO NOT REVERSE ALL THREE SETS OF LEADS.

NOTE: STANDARD ROTATION FOR D.C. GENERATOR IS CLOCKWISE

DIRECT CURRENT MOTORS

TO REVERSE THE ROTATION OF DIRECT CURRENT MOTORS:

DIRECT CURRENT MOTORS ARE REVERSED BY CHANGING THE DIRECTION OF THE FLOW OF THE CURRENT THROUGH THE ARMATURE, OR FIELD.

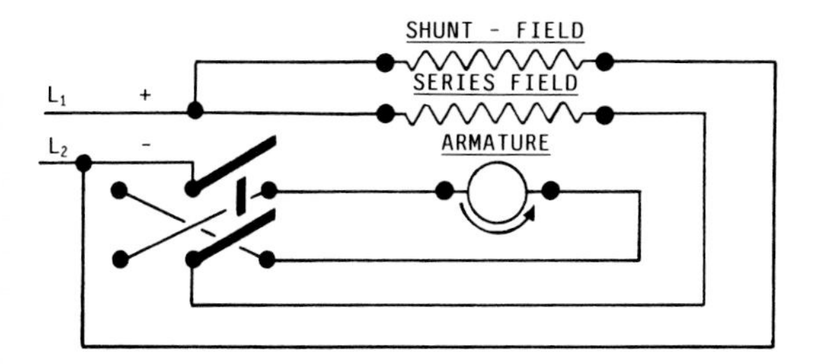

A COMPOUND D-C MOTOR CONNECTED TO A DOUBLE-POLE, DOUBLE THROW TRANSFER SWITCH.

TO CHANGE THE SPEED OF A D-C MOTOR:

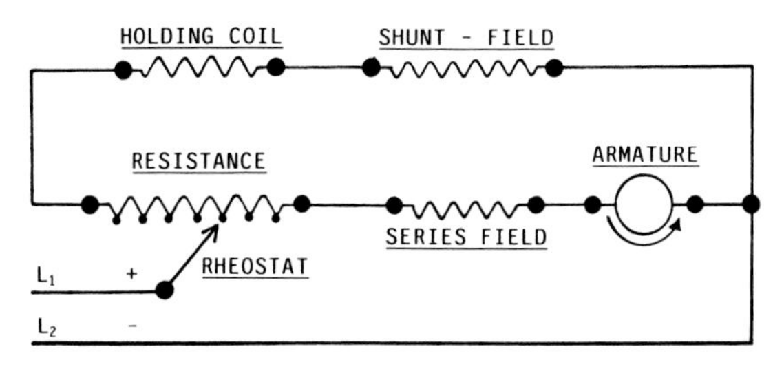

FULL-LOAD CURRENT IN AMPERES

SINGLE-PHASE ALTERNATING CURRENT MOTORS

HP	115V	200V	208V	230V
1/6	4.4	2.5	2.4	2.2
1/4	5.8	3.3	3.2	2.9
1/3	7.2	4.1	4	3.6
1/2	9.8	5.6	5.4	4.9
3/4	13.8	7.9	7.6	6.9
1	16	9.2	8.8	8
1-1/2	20	11.5	11	10
2	24	13.8	13.2	12
3	34	19.6	18.7	17
5	56	32.2	30.8	28
7-1/2	80	46	44	40
10	100	57.5	55	50

THE VOLTAGES LISTED ARE RATED MOTOR VOLTAGES. THE LISTED CURRENTS ARE FOR SYSTEM VOLTAGE RANGES OF 110 TO 120 AND 220 TO 240.

SINGLE-PHASE USING STANDARD THREE-PHASE STARTER

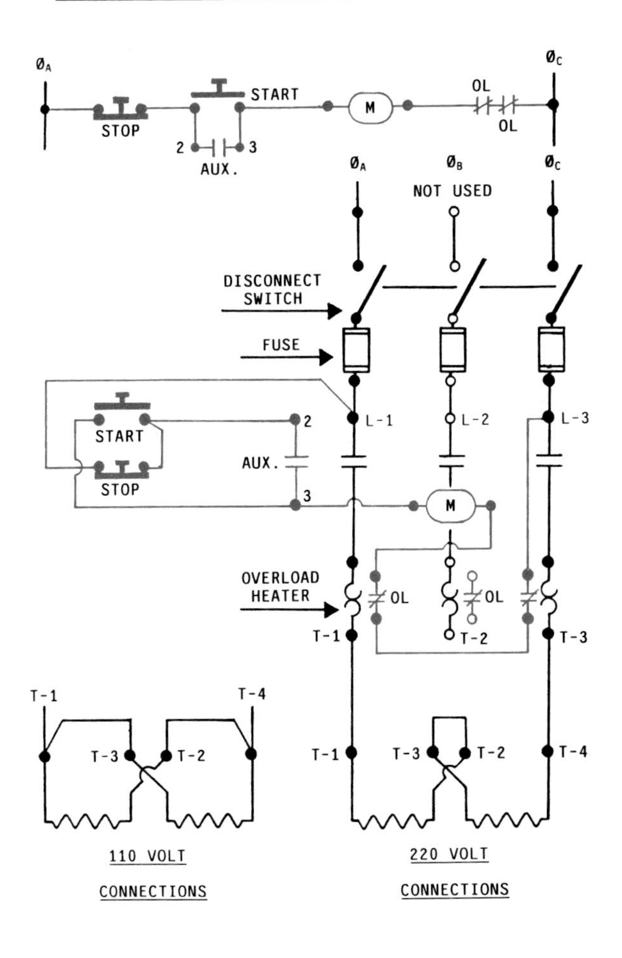

SINGLE PHASE MOTORS

SPLIT-PHASE---SQUIRREL CAGE---DUAL-VOLTAGE:

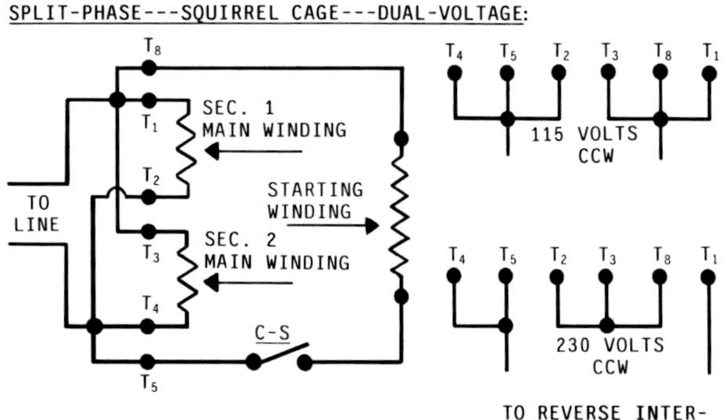

TO REVERSE INTER-
CHANGE 5 AND 8

CLASSES OF SINGLE PHASE MOTORS:

1. SPLIT-PHASE

   A. CAPACITOR-START
   B. REPULSION-START
   C. RESISTANCE-START
   D. SPLIT-CAPACITOR

2. COMMUTATOR

   A. REPULSION
   B. SERIES

TERMINAL COLOR MARKING:

T_1 BLUE T_3 ORANGE T_5 BLACK

T_2 WHITE T_4 YELLOW T_8 RED

NOTE: SPLIT-PHASE MOTORS ARE USUALLY FRACTIONAL HORSEPOWER. THE MAJORITY OF ELECTRIC MOTORS USED IN WASHING MACHINES, REFRIGERATORS, AND ETC. ARE OF THE SPLIT-PHASE TYPE.

TO CHANGE THE SPEED OF A SPLIT-PHASE MOTOR THE NUMBER OF POLES MUST BE CHANGED.

1. ADDITION OF RUNNING WINDING
2. TWO STARTING WINDINGS, AND TWO RUNNING WINDINGS
3. CONSEQUENT POLE CONNECTIONS.

SINGLE PHASE MOTORS

SPLIT-PHASE: SQUIRREL CAGE

A. RESISTANCE START:

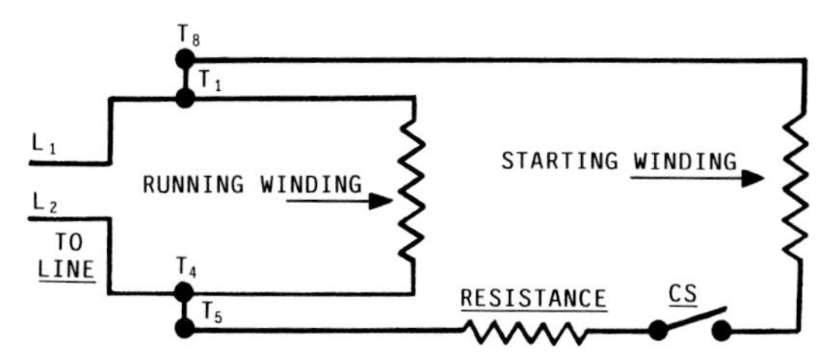

CENTRIFUGAL SWITCH (CS) OPENS AFTER REACHING 75% OF NORMAL SPEED.

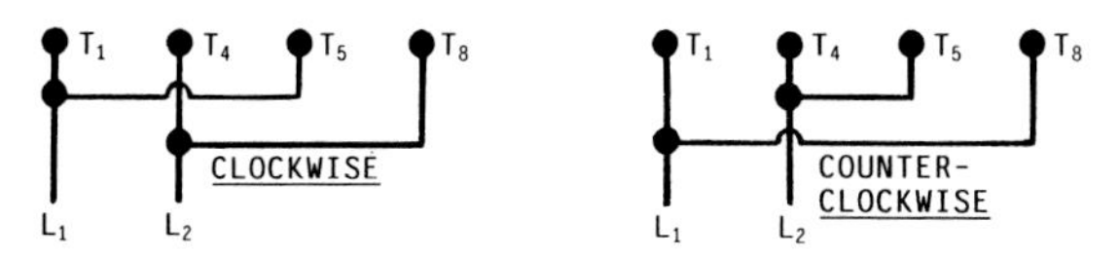

B. CAPACITOR START:

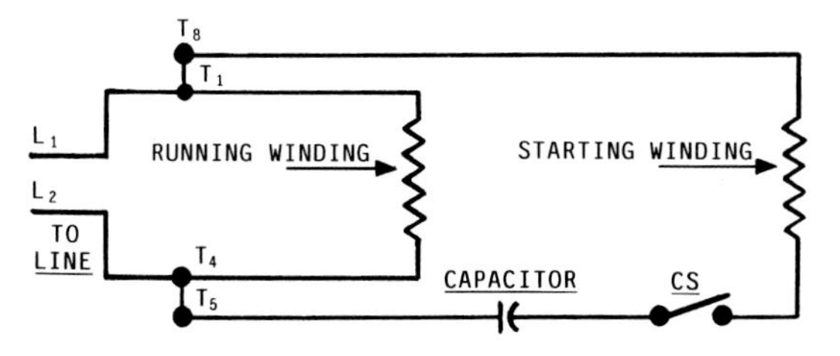

NOTE:

1. A RESISTANCE START MOTOR HAS A RESISTANCE CONNECTED IN SERIES WITH THE STARTING WINDING.
2. THE CAPACITOR START MOTOR IS EMPLOYED WHERE A HIGH STARTING TORQUE IS REQUIRED.

FULL-LOAD CURRENT

TWO-PHASE ALTERNATING-CURRENT MOTORS (4 WIRE)

INDUCTION TYPE SQUIRREL-CAGE AND WOUND ROTOR AMPERES					
HP	115V	230V	460V	575V	2300V
1/2	4	2	1	0.8	—
3/4	4.8	2.4	1.2	1.0	—
1	6.4	3.2	1.6	1.3	—
1-1/2	9	4.5	2.3	1.8	—
2	11.8	5.9	3	2.4	—
3	—	8.3	4.2	3.3	—
5	—	13.2	6.6	5.3	—
7-1/2	—	19	9	8	—
10	—	24	12	10	—
15	—	36	18	14	—
20	—	47	23	19	—
25	—	59	29	24	—
30	—	69	35	28	—
40	—	90	45	36	—
50	—	113	56	45	—
60	—	133	67	53	14
75	—	166	83	66	18
100	—	218	109	87	23
125	—	270	135	108	28
150	—	312	156	125	32
200	—	416	208	167	43

FOR 90 AND 80 PERCENT POWER FACTOR THE ABOVE FIGURES SHOULD BE MULTIPLIED BY 1.1 AND 1.25 RESPECTIVELY.

TWO-PHASE, FOUR WIRE

STANDARD THREE PHASE STARTER

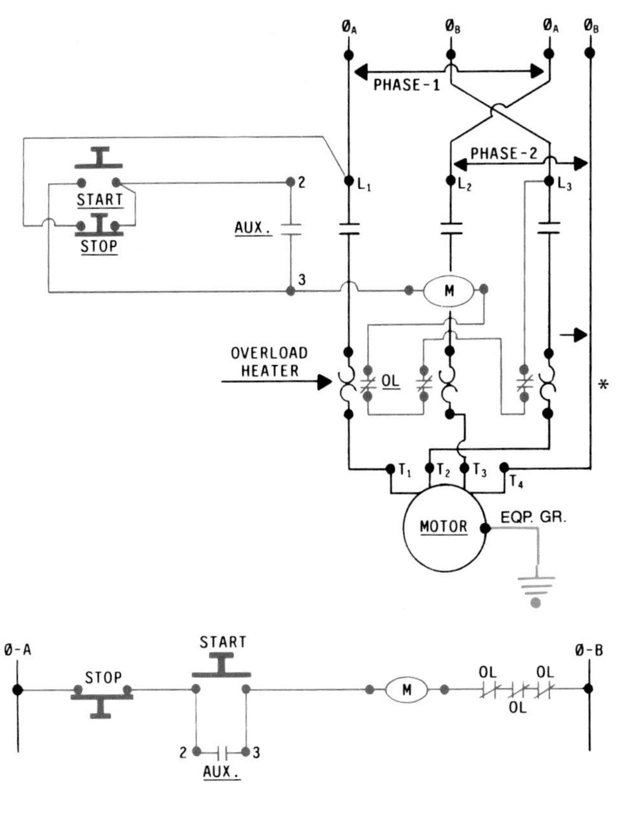

* NO HEATER OR HEATER OVERLOAD RELAY NECESSARY FOR T_4

TWO-PHASE MOTORS

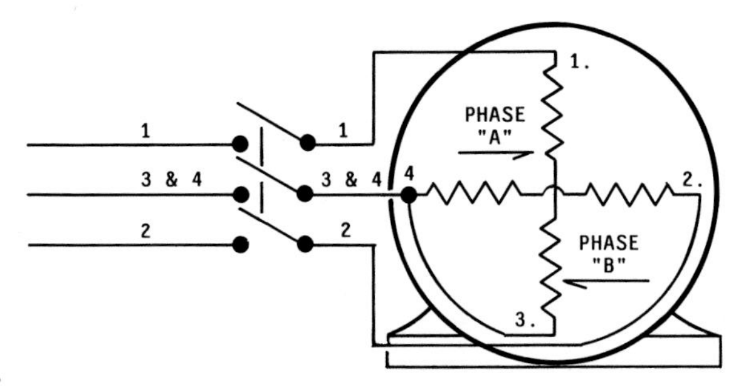

TWO PHASE --- THREE WIRE

TO REVERSE THE DIRECTION OF A TWO PHASE, THREE WIRE MOTOR INTERCHANGE THE TWO OUTSIDE MOTOR LEADS, 1 AND 2.

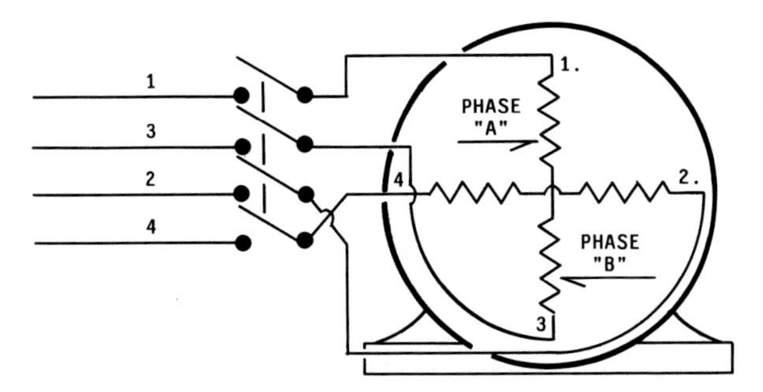

TWO PHASE ---- FOUR WIRE

TO REVERSE THE DIRECTION OF A TWO PHASE, FOUR WIRE MOTOR INTERCHANGE THE LEADS IN ONE PHASE.

FULL LOADS CURRENT
THREE-PHASE ALTERNATING CURRENT MOTORS

INDUCTION TYPE SQUIRREL-CAGE AND WOUND-ROTOR AMPERES								SYNCHRONOUS TYPE *UNITY POWER FACTOR AMPERES			
HP	115V	200V	208V	230V	460V	575V	2300V	230V	460V	575V	2300V
½	4	2.3	2.2	2	1	.8	-	-	-	-	-
¾	5.6	3.2	3.1	2.8	1.4	1.1	-	-	-	-	-
1	7.2	4.1	4.0	3.6	1.8	1.4	-	-	-	-	-
1½	10.4	6.0	5.7	5.2	2.6	2.1	-	-	-	-	-
2	13.6	7.8	7.5	6.8	3.4	2.7	-	-	-	-	-
3	-	11.0	10.6	9.6	4.8	3.9	-	-	-	-	-
5	-	17.5	16.7	15.2	7.6	6.1	-	-	-	-	-
7½	-	25.3	24.2	22.0	11.0	9.0	-	-	-	-	-
10	-	32.2	30.8	28.0	14.0	11.0	-	-	-	-	-
15	-	48.3	46.2	42.0	21.0	17.0	-	-	-	-	-
20	-	62.1	59.4	54.0	27.0	22.0	-	-	-	-	-
25	-	78.2	74.8	68.0	34.0	27.0	-	53.0	26.0	21.0	-
30	-	92.0	88.0	80.0	40.0	32.0	-	63.0	32.0	26.0	-
40	-	119.6	114.4	104.0	52.0	41.0	-	83.0	41.0	33.0	-
50	-	149.5	143.0	130.0	65.0	52.0	-	104.0	52.0	42.0	-
60	-	177.1	169.4	154.0	77.0	62.0	16.0	123.0	61.0	49.0	12.0
75	-	220.8	211.2	192.0	96.0	77.0	20.0	155.0	78.0	62.0	15.0
100	-	285.2	272.8	248.0	124.0	99.0	26.0	202.0	101.0	81.0	20.0
125	-	358.8	343.2	312.0	156.0	125.0	31.0	253.0	126.0	101.0	25.0
150	-	414.0	396.0	360.0	180.0	144.0	37.0	302.0	151.0	121.0	30.0
200	-	552.0	528.0	480.0	240.0	192.0	49.0	400.0	201.0	161.0	40.0

THE VOLTAGES LISTED ARE RATED MOTOR VOLTAGES. CORRESPONDING NOMINAL SYSTEM VOLTAGES ARE 110 TO 120, 220 TO 240, 440 TO 480, AND 550 TO 600 VOLTS.

*FOR 90 AND 80 PERCENT POWER FACTOR, THE ABOVE FIGURES SHALL BE MULTIPLIED BY 1.1 AND 1.25 RESPECTIVELY.

FULL-LOAD CURRENT AND OTHER DATA

THREE PHASE A.C. MOTORS

MOTOR HORSEPOWER		MOTOR AMPERE	SIZE BREAKER	SIZE STARTER	HEATER AMPERE	SIZE WIRE	SIZE CONDUIT
1/2	230V 460	2 1	15 15	00 00	3 2	12 12	3/4" 3/4
3/4	230 460	2.8 1.4	15 15	00 00	4 3	12 12	3/4 3/4
1	230 460	3.6 1.8	15 15	00 00	5 3	12 12	3/4 3/4
1½	230 460	5.2 2.6	15 15	00 00	8 4	12 12	3/4 3/4
2	230 460	6.8 3.4	15 15	0 00	9 5	12 12	3/4 3/4
3	230 460	9.6 4.8	15 15	0 0	12 7	12 12	3/4 3/4
5	230 460	15.2 7.6	15 15	1 0	20 10	12 12	3/4 3/4
7½	230 460	22 11	40 30	1 1	30 15	10 12	3/4 3/4
10	230 460	28 14	50 30	2 1	35 20	10 12	3/4 3/4
15	230 460	42 21	70 40	2 2	50 25	8 10	1 3/4
20	230 460	54 27	100 50	3 2	70 35	6 10	1 3/4
25	230 460	68 34	100 50	3 2	80 40	4 8	1½ 1
30	230 460	80 40	125 70	3 3	100 50	3 8	1½ 1
40	230 460	104 52	175 100	4 3	150 70	1 6	1½ 1
50	230 460	130 65	200 150	4 3	175 80	0 4	2 1½

★ OVERCURRENT DEVICE MAY HAVE TO BE INCREASED-SEE NEC CODE 430-152.

FULL-LOAD CURRENT AND OTHER DATA

THREE PHASE A.C. MOTORS

MOTOR HORSEPOWER		MOTOR AMPERE	SIZE BREAKER	SIZE STARTER	HEATER AMPERE	SIZE WIRE	SIZE CONDUIT
60	230V 460	154 77	250 200	5 4	200 100	000 3	2" 1½
75	230 460	192 96	300 200	5 4	250 125	0000 2	2½ 1½
100	230 460	248 124	400 200	5 4	300 175	350MCM 0	3 2
125	230 460	312 156	500 250	6 5	400 200	500MCM 000	3 2
150	230 460	360 180	600 300	6 5	450 225	700MCM 0000	4 2½

NOTE:

1. WIRE SIZE WILL VARY DEPENDING ON TYPE OF INSULATION.
2. THE PRECEEDING CALCULATIONS APPLY TO INDUCTION TYPE, SQUIRREL-CAGE, AND WOUND-ROTOR MOTORS ONLY.
3. THE VOLTAGES LISTED ARE RATED MOTOR VOLTAGES; CORRESPONDING NOMINAL SYSTEM VOLTAGES ARE 220V TO 240V, AND 440V TO 480V.
4. HERTZ: PREFERRED TERMINOLOGY FOR CYCLES PER SECOND.
5. FORM COIL: COIL MADE WITH RECTANGULAR OR SQUARE WIRE.
6. MUSH COIL: COIL MADE WITH ROUND WIRE.
7. SLIP: PERCENTAGE DIFFERENCE BETWEEN SYNCHRONOUS AND OPERATING SPEEDS.
8. SYCHRONOUS SPEED: MAXIMUM SPEED FOR A.C. MOTORS OR (FREQUENCY × 120) / POLES
9. FULL LOAD SPEED: SPEED AT WHICH RATED HORSEPOWER IS DEVELOPED.
10. POLES: NUMBER OF MAGNETIC POLES SET UP INSIDE THE MOTOR BY THE PLACEMENT AND CONNECTION OF THE WINDINGS.

THREE PHASE A.C. MOTOR WINDINGS AND CONNECTIONS

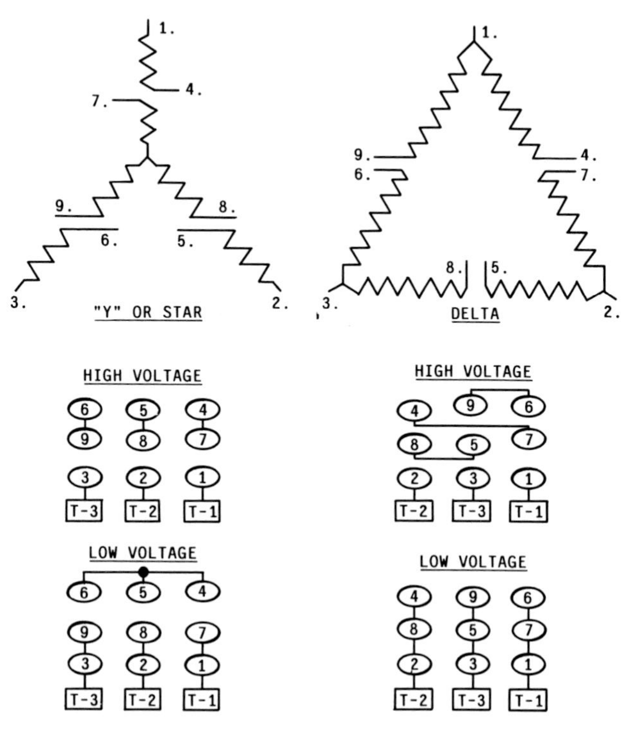

NOTE:

- THE MOST IMPORTANT PART OF ANY MOTOR IS THE NAME-PLATE. CHECK THE DATA GIVEN ON THE PLATE BEFORE MAKING THE CONNECTIONS.
- TO CHANGE ROTATION DIRECTION OF 3 PHASE MOTOR, SWAP ANY 2 T-LEADS.

THREE WIRE STOP-START STATION

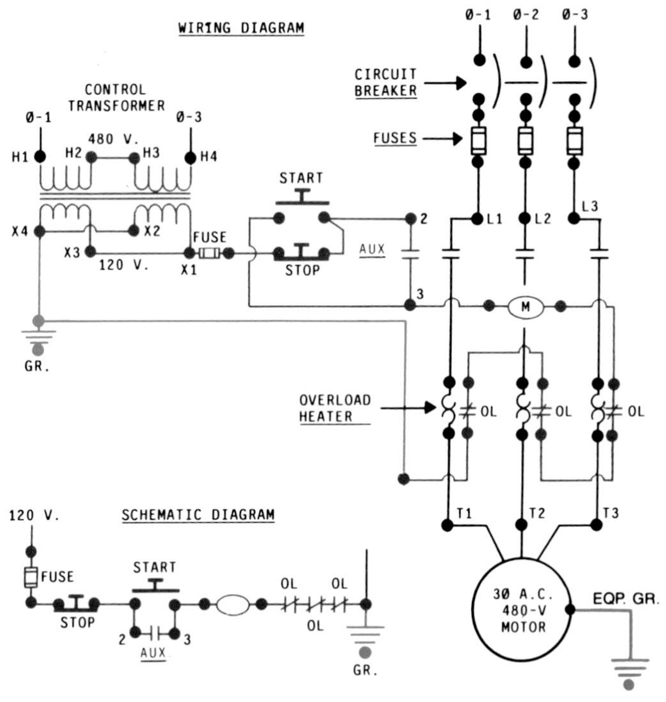

NOTE: CONTROLS AND MOTOR ARE OF DIFFERENT VOLTAGES.

TWO THREE WIRE STOP-START STATIONS

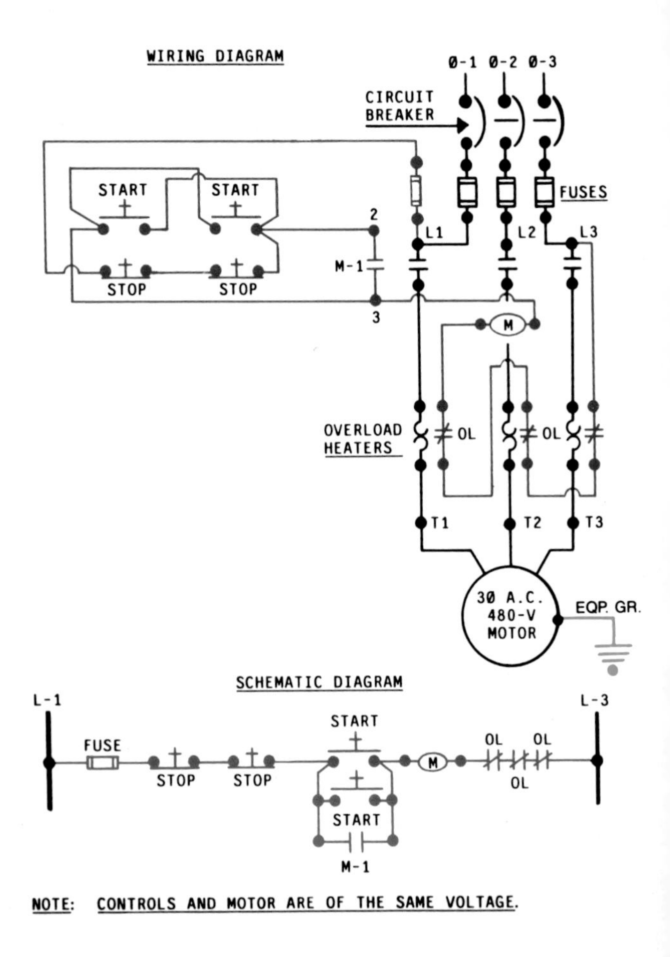

NOTE: CONTROLS AND MOTOR ARE OF THE SAME VOLTAGE.

HAND OFF AUTOMATIC CONTROL

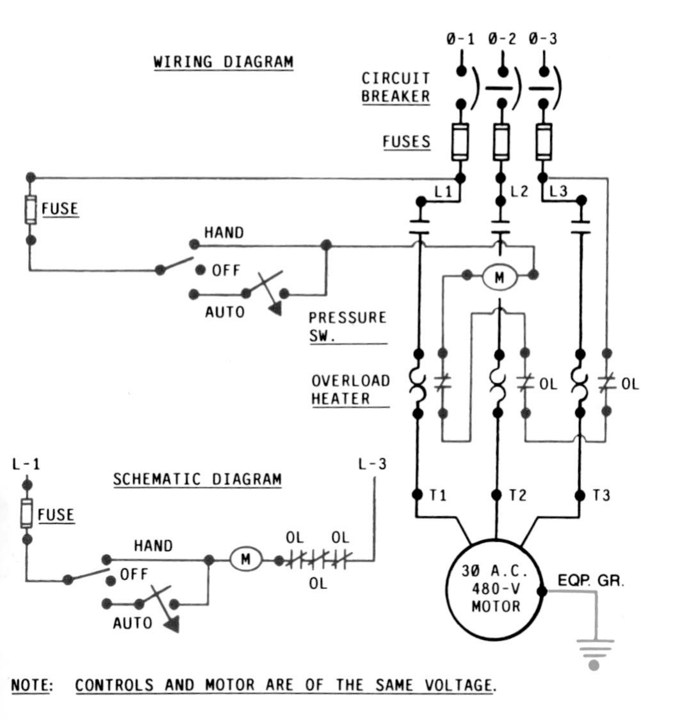

NOTE: CONTROLS AND MOTOR ARE OF THE SAME VOLTAGE.

JOGGING WITH CONTROL RELAY

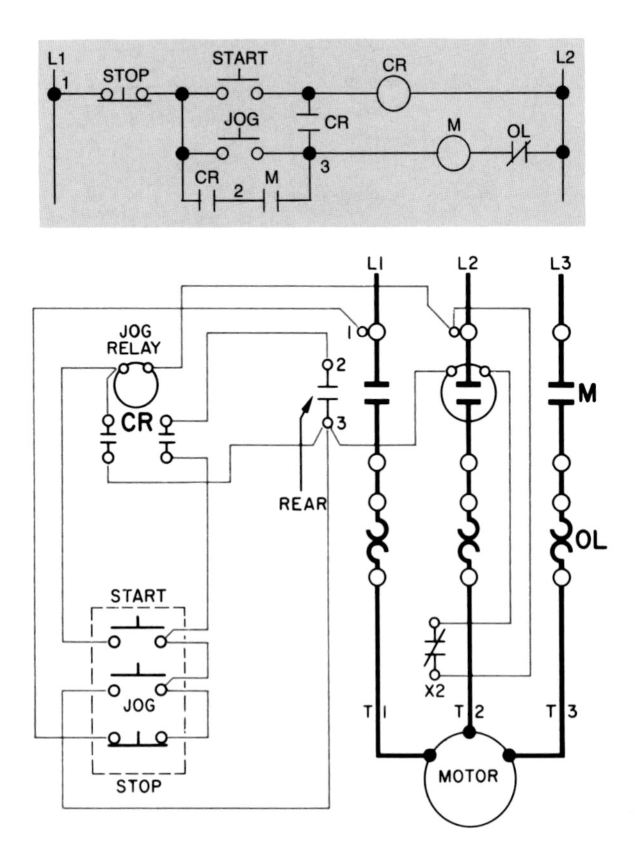

JOGGING CIRCUITS ARE USED WHEN MACHINES MUST BE OPERATED MOMENTAR FOR INCHING, (AS IN SET UP OR MAINTENANCE). THE JOG CIRCUIT ALLOWS STARTER TO BE ENERGIZED ONLY AS LONG AS THE JOG BUTTON IS DEPRESS

TRANSFORMER CALCULATIONS

TO BETTER UNDERSTAND THE FOLLOWING FORMULAS REVIEW THE RULE OF TRANSPOSITION IN EQUATIONS.

A MULTIPLIER MAY BE REMOVED FROM ONE SIDE OF AN EQUATION BY MAKING IT A DIVISOR ON THE OTHER SIDE, OR A DIVISOR MAY BE REMOVED FROM ONE SIDE OF AN EQUATION BY MAKING IT A MULTIPLIER ON THE OTHER SIDE.

1. VOLTAGE AND CURRENT: PRIMARY (p) AND SECONDARY (s)

POWER (p) = POWER (s) OR $Ep \times Ip = Es \times Is$

A. $Ep = \frac{Es \times Is}{Ip}$

B. $Ip = \frac{Es \times Is}{Ep}$

C. $\frac{Ep \times Ip}{Es} = Is$

D. $\frac{Ep \times Ip}{Is} = Es$

2. VOLTAGE AND TURNS IN COIL:

VOLTAGE (p) × TURNS (s) = VOLTAGE (s) × TURNS (p)

OR

$$Ep \times Ts = Es \times Tp$$

A. $EP = \frac{Es \times Tp}{Ts}$

B. $Ts = \frac{Es \times Tp}{Ep}$

C. $\frac{Ep \times Ts}{Es} = Tp$

D. $\frac{Ep \times Ts}{Tp} = Es$

3. AMPERES AND TURNS IN COIL:

AMPERES (p) × TURNS (p) = AMPERES (s) × TURNS (s)

OR

$$Ip \times Tp = Is \times Ts$$

A. $Ip = \frac{Is \times Ts}{Tp}$

B. $Tp = \frac{Is \times Ts}{Ip}$

C. $\frac{Ip \times Tp}{Is} = Ts$

D. $\frac{Ip \times Tp}{Ts} = Is$

VOLTAGE DROP CALCULATIONS

INDUCTANCE NEGLIGIBLE

V = DROP IN CIRCUIT VOLTAGE
R = RESISTANCE PER FT. OF CONDUCTOR (OHMS / FT.)
I = CURRENT IN CONDUCTOR (AMPERES)
L = ONE-WAY LENGTH OF CIRCUIT (FT.)
D = CROSS SECTION AREA OF CONDUCTOR (CIRCULAR MILS)
K = RESISTIVITY OF CONDUCTOR @75°C

A. K = 12.9 FOR COPPER CONDUCTORS

B. K = 21.2 FOR ALUMINUM CONDUCTORS

1. TWO-WIRE SINGLE PHASE CIRCUITS:

$$V = \frac{2K \times L \times I}{D}$$

2. THREE-WIRE SINGLE PHASE CIRCUITS:

$$V = \frac{2K \times L \times I}{D}$$

3. THREE-WIRE THREE PHASE CIRCUITS:

$$V = \frac{2K \times L \times I}{D} \times 0.866$$

4. FOUR-WIRE THREE PHASE BALANCED CIRCUITS:

$$V = \frac{2K \times L \times I}{D} \times \frac{1}{2}$$

NOTE:
1. FOR LIGHTING LOADS: VOLTAGE DROP BETWEEN ONE OUTSIDE CONDUCTOR AND NEUTRAL EQUALS ONE-HALF OF DROP CALCULATED BY FORMULA FOR TWO-WIRE CIRCUITS.
2. FOR MOTOR LEADS: VOLTAGE DROP BETWEEN ANY TWO OUTSIDE CONDUCTORS EQUALS 0.866 TIMES DROP DETERMINED BY FORMULA FOR TWO-WIRE CIRCUITS.

SINGLE-PHASE TRANSFORMER CONNECTIONS

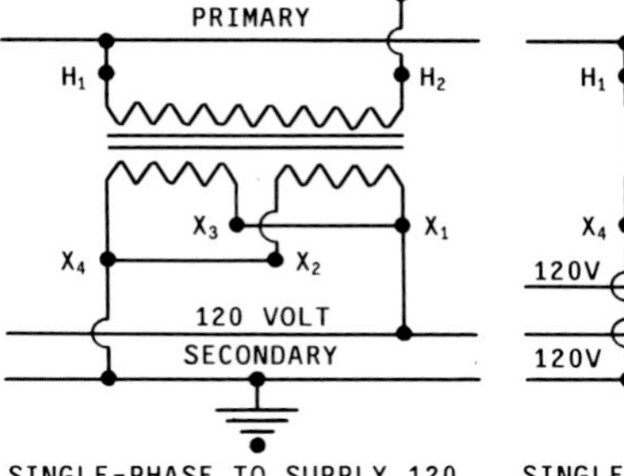

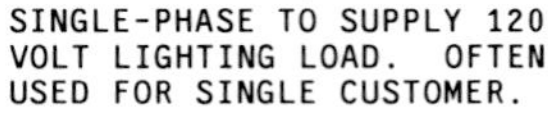

SINGLE-PHASE TO SUPPLY 120 VOLT LIGHTING LOAD. OFTEN USED FOR SINGLE CUSTOMER.

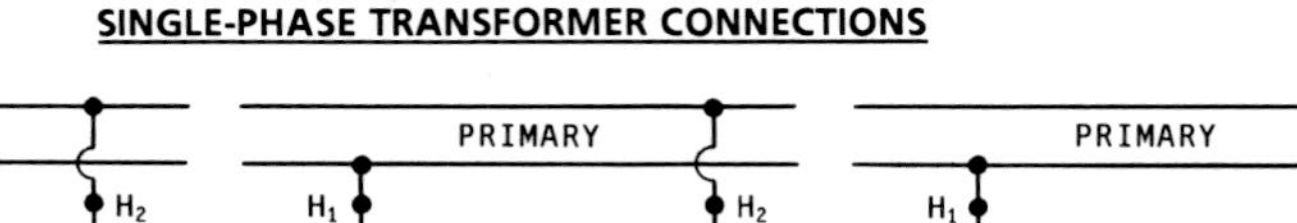

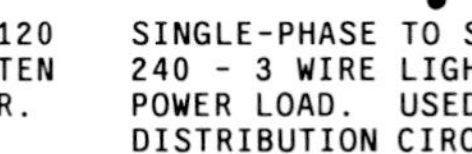
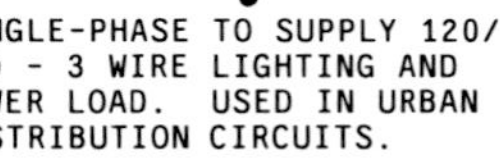

SINGLE-PHASE TO SUPPLY 120/240 - 3 WIRE LIGHTING AND POWER LOAD. USED IN URBAN DISTRIBUTION CIRCUITS.

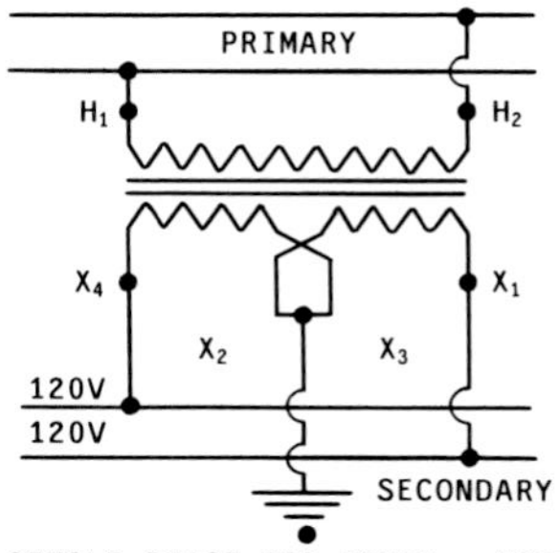

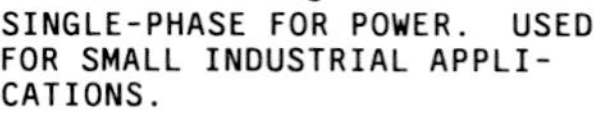

SINGLE-PHASE FOR POWER. USED FOR SMALL INDUSTRIAL APPLICATIONS.

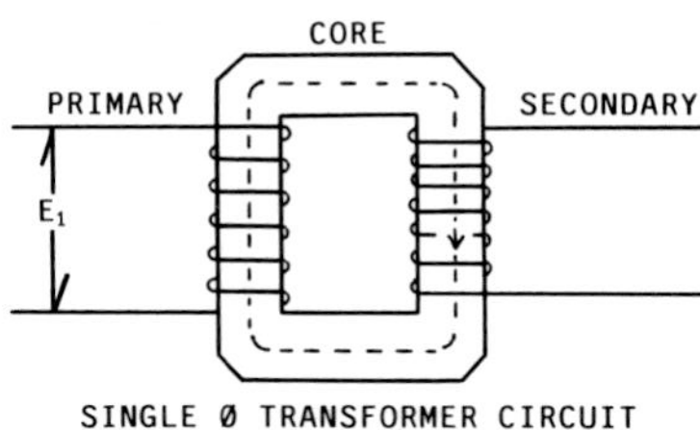

SINGLE Ø TRANSFORMER CIRCUIT

A TRANSFORMER IS A STATIONARY INDUCTION DEVICE FOR TRANSFERRING ELECTRICAL ENERGY FROM ONE CIRCUIT TO ANOTHER WITHOUT CHANGE OF FREQUENCY. A TRANSFORMER CONSISTS OF TWO COILS OR WINDINGS WOUND UPON A MAGNETIC CORE OF SOFT IRON LAMINATIONS, AND INSULATED FROM ONE ANOTHER.

BUCK AND BOOST TRANSFORMER CONNECTONS

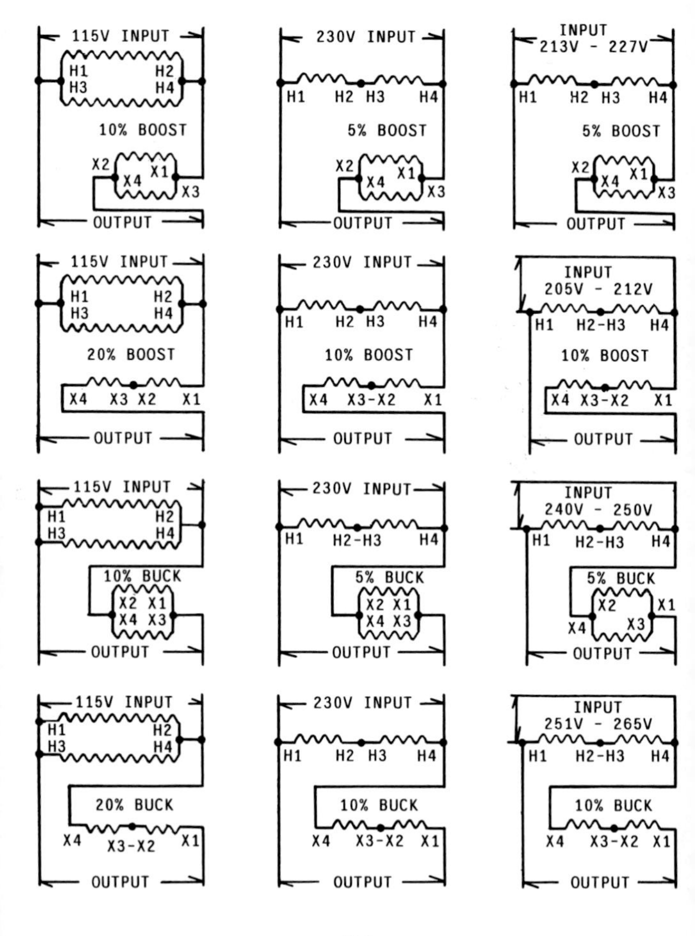

FULL LOAD CURRENTS

THREE-PHASE TRANSFORMERS

VOLTAGE (LINE TO LINE)

KVA RATING	208	240	480	2400	4160
3	8.3	7.2	3.6	.72	.415
6	16.6	14.4	7.2	1.44	.83
9	25	21.6	10.8	2.16	1.25
15	41.6	36.0	18.0	3.6	2.1
30	83	72	36	7.2	4.15
45	125	108	54	10.8	6.25
75	208	180	90	18	10.4
100	278	241	120	24	13.9
150	416	360	180	36	20.8
225	625	542	271	54	31.2
300	830	720	360	72	41.5
500	1390	1200	600	120	69.4
750	2080	1800	900	180	104
1000	2775	2400	1200	240	139
1500	4150	3600	1800	360	208
2000	5550	4800	2400	480	277
2500	6950	6000	3000	600	346
5000	13900	12000	6000	1200	694
7500	20800	18000	9000	1800	1040
10000	27750	24000	12000	2400	1386

KVA = E × I × 1.73 / 1000

SINGLE - PHASE TRANSFORMERS

VOLTAGE

KVA RATING	120	208	240	480	2400
1	8.34	4.8	4.16	2.08	.42
3	25	14.4	12.5	6.25	1.25
5	41.7	24.0	20.8	10.4	2.08
7.5	62.5	36.1	31.2	15.6	3.12
10	83.4	48	41.6	20.8	4.16
15	125	72	62.5	31.2	6.25
25	208	120	104	52	10.4
37.5	312	180	156	78	15.6
50	417	240	208	104	20.8
75	625	361	312	156	31.2
100	834	480	416	208	41.6
125	1042	600	520	260	52.0
167.5	1396	805	698	349	70.0
200	1666	960	833	416	83.3
250	2080	1200	1040	520	104
333	2776	1600	1388	694	139
500	4170	2400	2080	1040	208

KVA = E × I / 1000

THREE PHASE CONNECTIONS

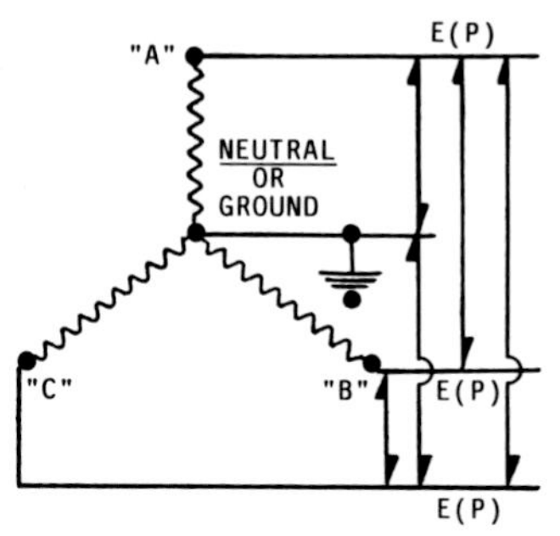

STAR

VOLTAGE FROM "A", "B", OR "C" TO GROUND = E(G)

VOLTAGE BETWEEN A-B, A-C, OR B-C = E(P)

$E(P) = E(G) \times 1.73$

$E(G) = E(P) / 1.73$

$POWER = 3 \times E(G) \times I \times \cos\theta$

DELTA

I(W) = CURRENT OF WINDING
I(P) = CURRENT OF PHASE

DELTA "E" = STAR "E" × 1.73
STAR "E" = DELTA "E" / 1.73

STAR "I" = DELTA "I" × 1.73
DELTA "I" = STAR "I" / 1.73

$POWER = 3 \times E(W) \times \cos\theta$

$I(P) = I(W) \times 1.73$

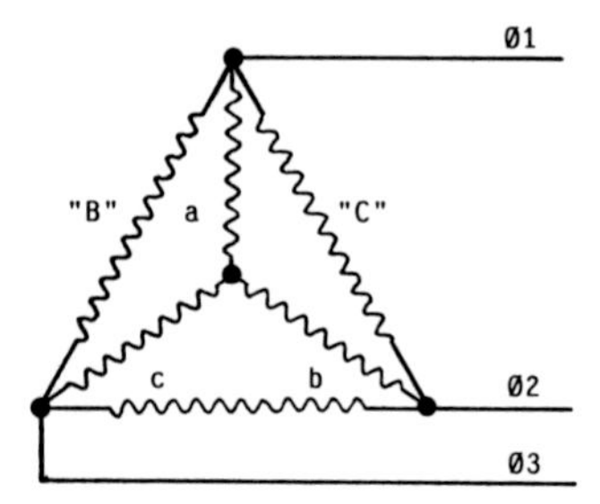

EQUIVALENT WYE-DELTA NETWORKS

$K(1) = A + B + C$
$K(2) = ab + ac + bc$

$$a = \frac{B \times C}{K(1)} \qquad A = \frac{K(2)}{a}$$

$$b = \frac{A \times C}{K(1)} \qquad B = \frac{K(2)}{b}$$

$$c = \frac{A \times B}{K(1)} \qquad C = \frac{K(2)}{c}$$

THREE-PHASE STANDARD PHASE ROTATION

TRANSFORMERS

STAR-DELTA

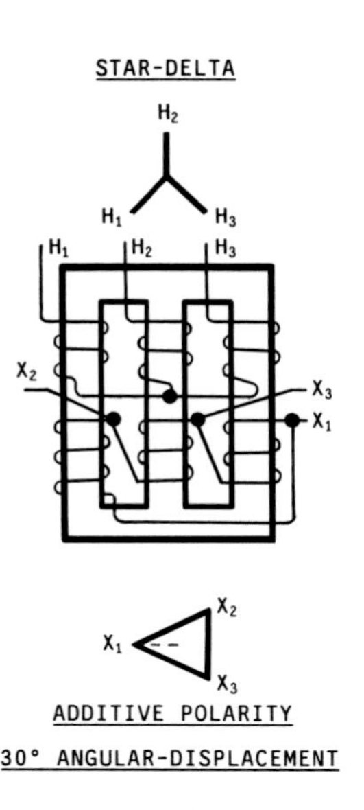

ADDITIVE POLARITY

30° ANGULAR-DISPLACEMENT

STAR-STAR

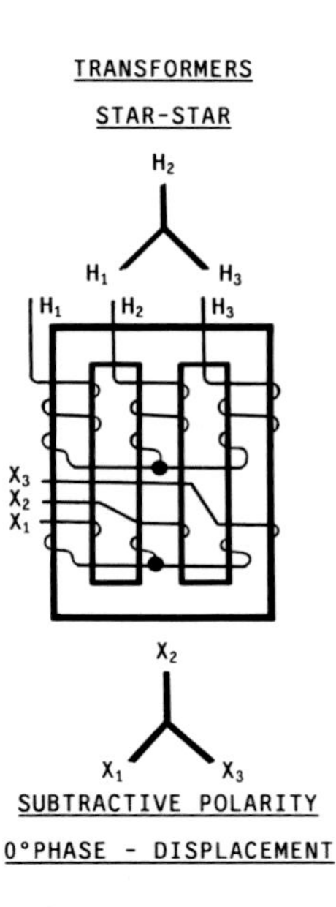

SUBTRACTIVE POLARITY

0°PHASE - DISPLACEMENT

DELTA-DELTA

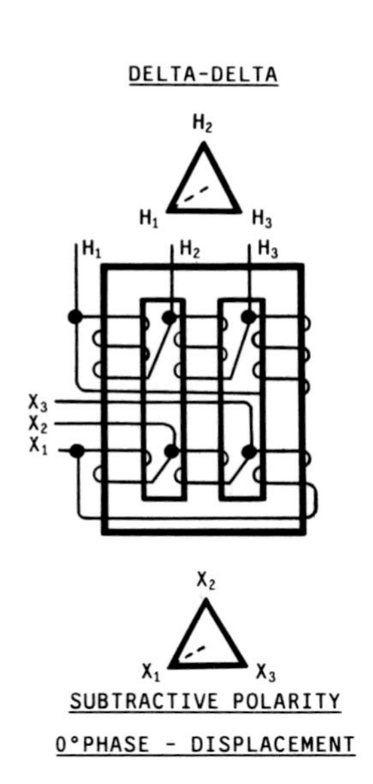

SUBTRACTIVE POLARITY

0°PHASE - DISPLACEMENT

TRANSFORMER CONNECTIONS

SERIES CONNECTION OF LOW VOLTAGE WINDINGS

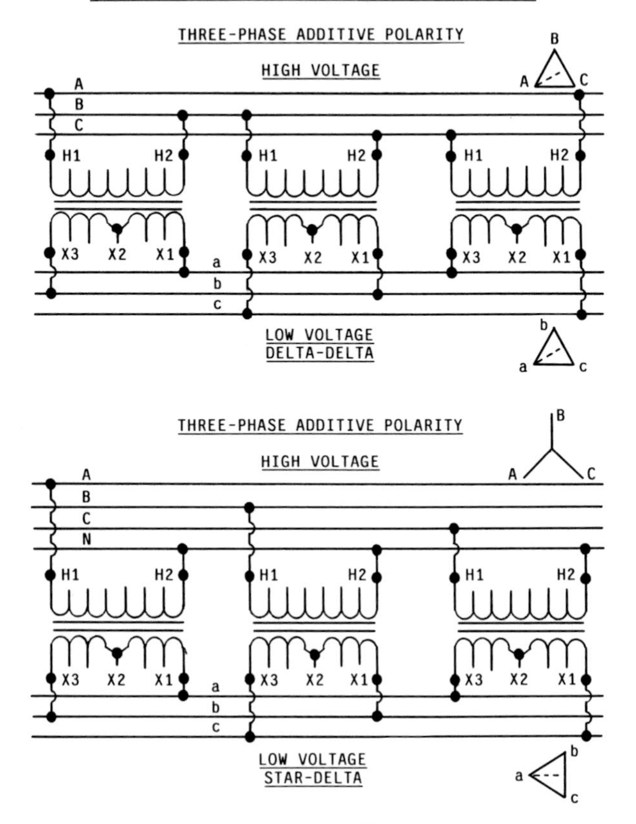

NOTE: SINGLE-PHASE TRANSFORMERS SHOULD BE THOROUGHLY CHECKED FOR IMPEDANCE. POLARITY, AND VOLTAGE RATIO BEFORE INSTALLATION.

TRANSFORMER CONNECTIONS

SERIES CONNECTION OF LOW VOLTAGE WINDINGS

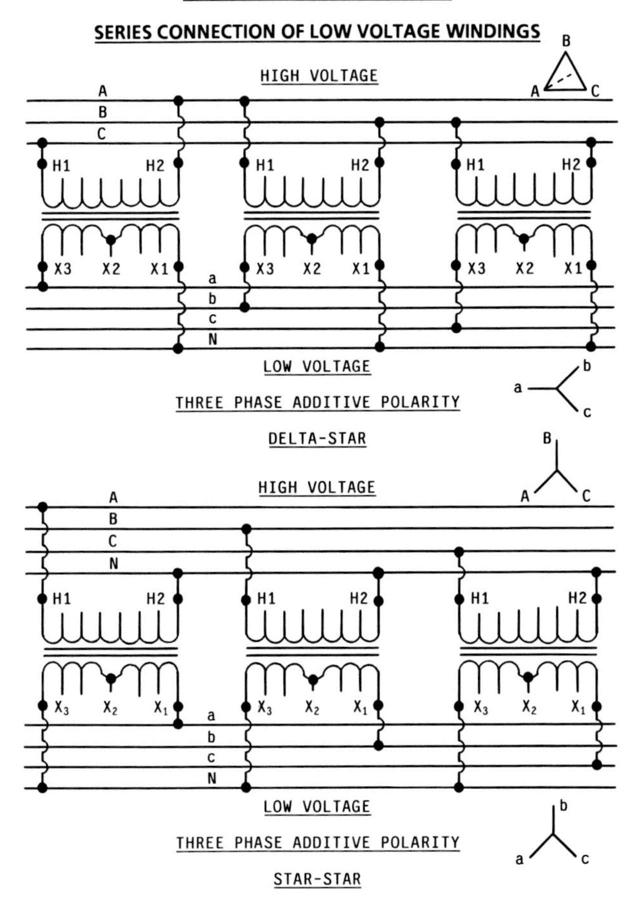

NOTE: FOR ADDITIVE POLARITY THE H-1 AND THE X-1 BUSHINGS ARE DIAGONALLY OPPOSITE EACH OTHER.

TRANSFORMER CONNECTIONS

SERIES CONNECTION OF LOW VOLTAGE WINDINGS

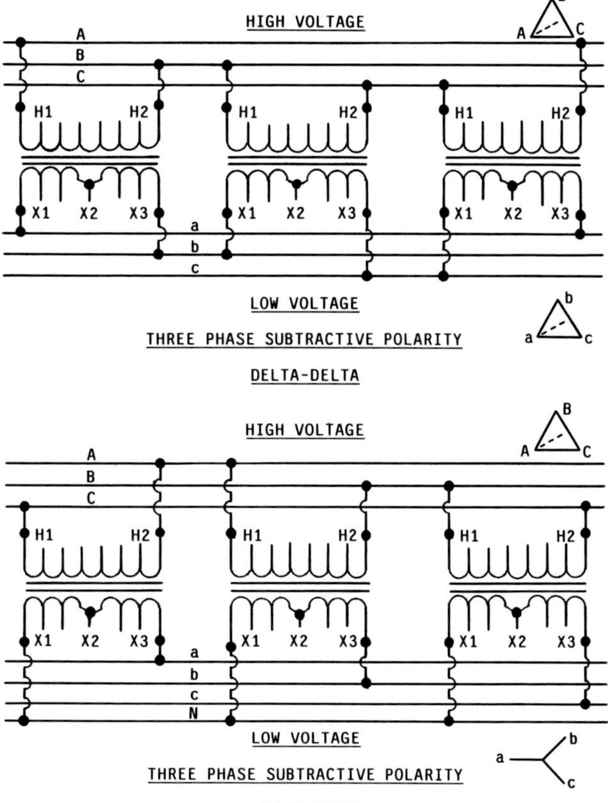

NOTE: FOR SUBTRACTIVE POLARITY THE H-1 AND THE X-1 BUSHINGS ARE DIRECTLY OPPOSITE EACH OTHER.

TRANSFORMER CONNECTIONS

TWO PHASE-FOUR WIRE

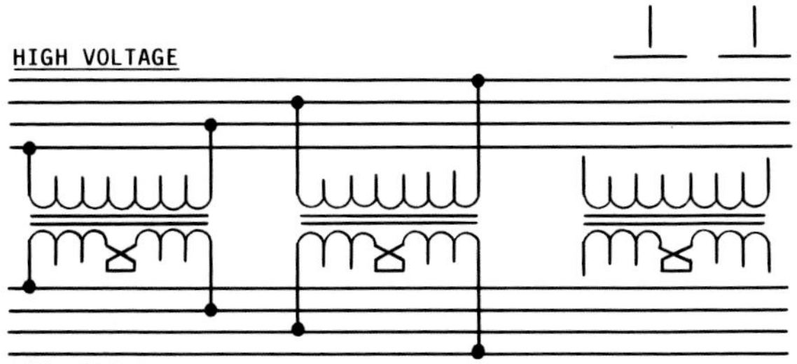

TWO-PHASE FOUR-WIRE IS TRANSFORMED TO TWO-PHASE FOUR-WIRE OF A DIFFERENT VOLTAGE WITH NO CONNECTION BETWEEN THE TWO PHASES.

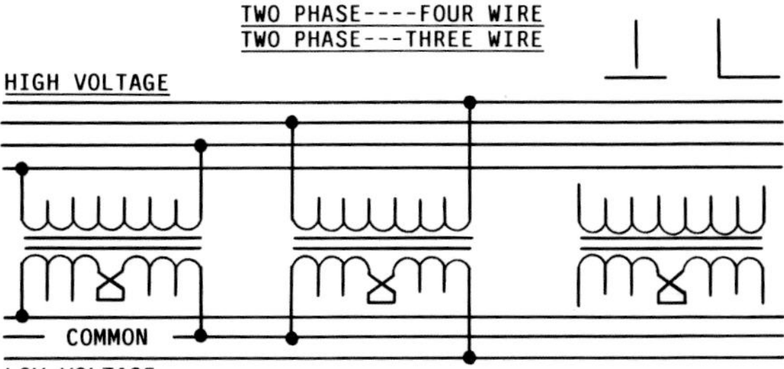

THE TWO PHASES ON THE LOW VOLTAGE SIDE ARE ELECTRICALLY CONNECTED. WITH BALANCED LOAD THE CURRENT IN THE COMMON WIRE IS 1.41 GREATER THAN THE CURRENT IN EITHER OF OUTSIDE WIRES.

TRANSFORMER CONNECTIONS

TWO PHASE-THREE WIRE

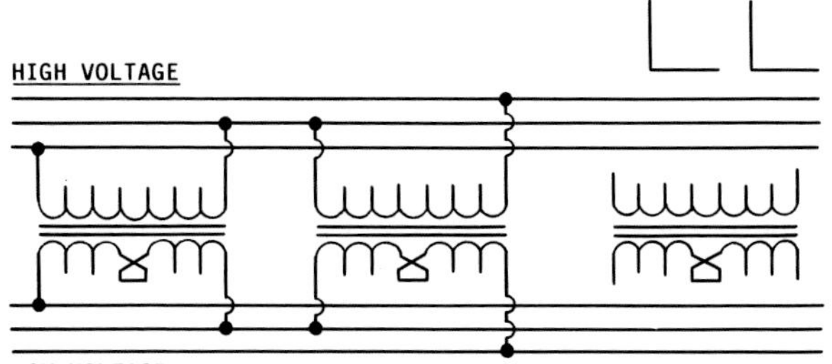

BOTH PHASES ARE ELECTRICALLY CONNECTED BY THE COMMON. THE COMMON IS SOMETIMES GROUNDED. WITH THE BALANCED LOAD THE CURRENT IN THE COMMON IS 1.41 TIMES THAT IN THE OUTSIDE LEGS.

THREE PHASE-OPEN DELTA

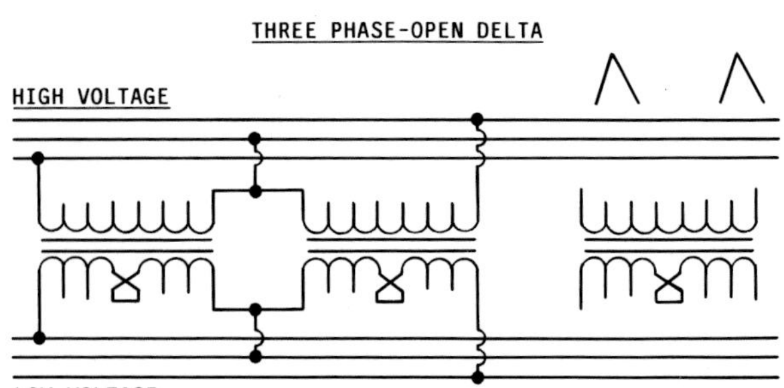

IN THIS OPEN DELTA CONNECTION THE UNITS WILL TRANSFORM 86% OF THEIR RATING.

IT IS NOT NECESSARY THAT THE IMPEDANCE CHARACTERISTICS BE IDENTICAL AS WITH THREE UNIT BANKS.

REGULATION OF OPEN-DELTA BANK IS NOT AS GOOD AS A CLOSED-DELTA BANK.

MISCELLANEOUS WIRING DIAGRAMS

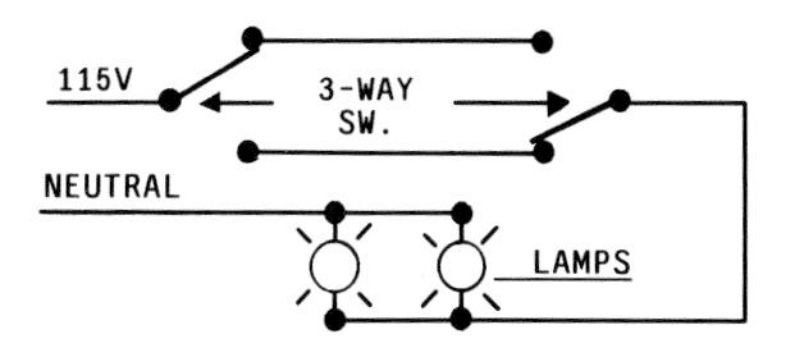

TWO 3-WAY SWITCHES

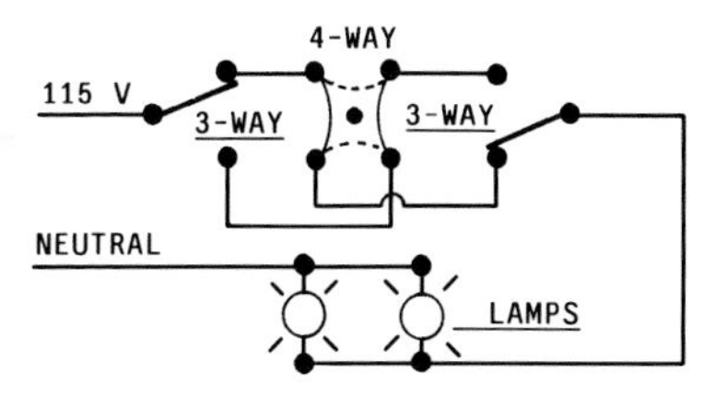

TWO 3-WAY SWITCHES
ONE 4-WAY SWITCH

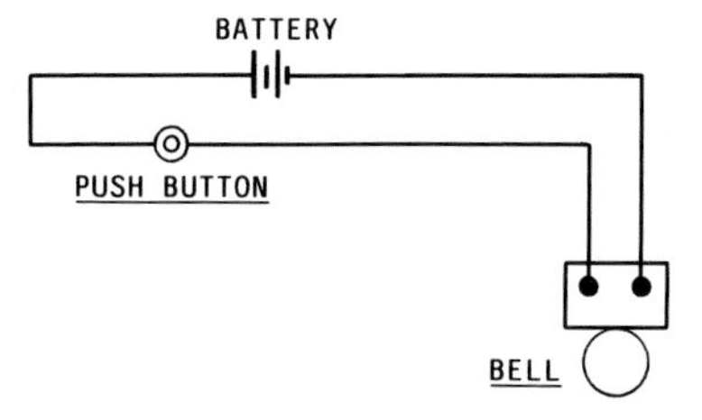

BELL CIRCUIT

MISCELLANEOUS WIRING DIAGRAMS

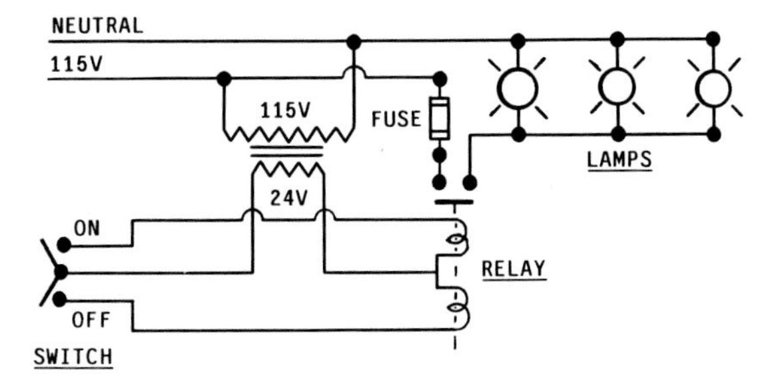

REMOTE CONTROL CIRCUIT - ONE RELAY AND ONE SWITCH

SUPPORTS FOR RIGID METAL CONDUIT

CONDUIT SIZE	DISTANCE BETWEEN SUPPORTS
1/2" - 3/4"	10 FEET
1"	12 FEET
1-1/4" - 1-1/2"	14 FEET
2" - 2-1/2"	16 FEET
3" AND LARGER	20 FEET

SUPPORT OF RIGID NONMETALLIC CONDUIT

CONDUIT SIZE (Inches)	MAXIMUM SPACING BETWEEN SUPPORTS (Feet)
1/2" - 1"	3 FEET
1-1/4" - 2"	5 FEET
2-1/2" - 3"	6 FEET
3-1/2" - 5"	7 FEET
6"	8 FEET

For SI units: (Supports) one foot = 0.3048 meter.

CONDUCTOR PROPERTIES

Size AWG/ kcmil	Area Cir. Mils	Conductors				DC Resistance at 75°C (167°F)		
		Stranding		Overall		Copper		Aluminum
		Quantity	Diam. In.	Diam. In.	Area In.^{2}	Uncoated ohm/MFT	Coated ohm/MFT	ohm/ MFT
18	1620	1	—	0.040	0.001	7.77	8.08	12.8
18	1620	7	0.015	0.046	0.002	7.95	8.45	13.1
16	2580	1	—	0.051	0.002	4.89	5.08	8.05
16	2580	7	0.019	0.058	0.003	4.99	5.29	8.21
14	4110	1	—	0.064	0.003	3.07	3.19	5.06
14	4110	7	0.024	0.073	0.004	3.14	3.26	5.17
12	6530	1	—	0.081	0.005	1.93	2.01	3.18
12	6530	7	0.030	0.092	0.006	1.98	2.05	3.25
10	10380	1	—	0.102	0.008	1.21	1.26	2.00
10	10380	7	0.038	0.116	0.011	1.24	1.29	2.04
8	16510	1	—	0.128	0.013	0.764	0.786	1.26
8	16510	7	0.049	0.146	0.017	0.778	0.809	1.28
6	26240	7	0.061	0.184	0.027	0.491	0.510	0.808
4	41740	7	0.077	0.232	0.042	0.308	0.321	0.508
3	52620	7	0.087	0.260	0.053	0.245	0.254	0.403
2	66360	7	0.097	0.292	0.067	0.194	0.201	0.319
1	83690	19	0.066	0.332	0.087	0.154	0.160	0.253
1/0	105600	19	0.074	0.373	0.109	0.122	0.127	0.201
2/0	133100	19	0.084	0.419	0.138	0.0967	0.101	0.159
3/0	167800	19	0.094	0.470	0.173	0.0766	0.0797	0.126
4/0	211600	19	0.106	0.528	0.219	0.0608	0.0626	0.100
250	—	37	0.082	0.575	0.260	0.0515	0.0535	0.0847
300	—	37	0.090	0.630	0.312	0.0429	0.0446	0.0707
350	—	37	0.097	0.681	0.364	0.0367	0.0382	0.0605
400	—	37	0.104	0.728	0.416	0.0321	0.0331	0.0529
500	—	37	0.116	0.813	0.519	0.0258	0.0265	0.0424
600	—	61	0.099	0.893	0.626	0.0214	0.0223	0.0353
700	—	61	0.107	0.964	0.730	0.0184	0.0189	0.0303
750	—	61	0.111	0.998	0.782	0.0171	0.0176	0.0282
800	—	61	0.114	1.03	0.834	0.0161	0.0166	0.0265
900	—	61	0.122	1.09	0.940	0.0143	0.0147	0.0235
1000	—	61	0.128	1.15	1.04	0.0129	0.0132	0.0212
1250	—	91	0.117	1.29	1.30	0.0103	0.0106	0.0169
1500	—	91	0.128	1.41	1.57	0.00858	0.00883	0.0141
1750	—	127	0.117	1.52	1.83	0.00735	0.00756	0.0121
2000	—	127	0.126	1.63	2.09	0.00643	0.00662	0.0106

COPPER

AMPACITIES OF SINGLE INSULATED CONDUCTOR RATED 0-2,000 VOLTS IN FREE AIR							
BASED ON AMBIENT AIR TEMPERATURE OF 30°C (86°F)				BASED ON AMBIENT TEMPERATURE OF 40°C (104°F)			
SIZE	60°C (140°F)	75°C (167°F)	90°C (194°F)	150°C (302°F)	200°C (392°F)	250°C (482°F)	BARE
AWG kcmil	TW* UF*	FEPW* THWN* XHHW* THHW* RHW* THW* RH* ZW*	TA,TBS,SA SIS,FEP*,RHH* FEPB*,MI,XHH RHW-2,THHN* THHW*,XHHW* THW-2,USE-2 THWN-2,ZW-2 XHHW-2	Z	FEP FEPB PFA	PFAH TFE NICKEL OR NICKEL COATED COPPER	
14	25*	30*	35*	46	54	59	30
12	30*	35*	40*	60	68	78	35
10	40*	50*	55*	80	90	107	50
8	60	70	80	106	124	142	70
6	80	95	105	155	165	205	95
4	105	125	140	190	220	278	125
3	120	145	165	214	252	327	150
2	140	170	190	255	293	381	175
1	165	195	220	293	344	440	200
0	195	230	260	339	399	532	235
00	225	265	300	390	467	591	275
000	260	310	350	451	546	708	320
0000	300	360	405	529	629	830	370
250	340	405	455	–	–	–	415
300	375	445	505	–	–	–	460
350	420	505	570	–	–	–	520
400	455	545	615	–	–	–	560
500	515	620	700	–	–	–	635
600	575	690	780	–	–	–	710
700	630	755	855	–	–	–	780
750	655	785	885	–	–	–	805
800	680	815	920	–	–	–	835
900	730	870	985	–	–	–	865
1000	780	935	1055	–	–	–	895
1250	890	1065	1200	–	–	–	–
1500	980	1175	1325	–	–	–	1205
1750	1070	1280	1445	–	–	–	–
2000	1155	1385	1560	–	–	–	1420

OVERCURRENT PROTECTION FOR CONDUCTOR TYPES MARKED (*) WILL NOT EXCEED 15 AMPERES FOR SIZE 14 AWG, 20 AMPERES FOR SIZE 12 AWG, AND 30 AMPERES FOR SIZE 10 AWG. (–) FOR DRY LOCATIONS ONLY. SEE 75°C COLUMN FOR WET LOCATIONS.

FOR AMBIENT TEMPERATURES OTHER THAN NOTED ABOVE, SEE N.E.C. TABLE 310-17 AND 310-19.

COPPER

AMPACITIES OF NOT MORE THAN THREE SINGLE INSULATED CONDUCTORS RATED 0-2,000 VOLTS						
IN RACEWAY, CABLE, OR EARTH, BASED ON AMBIENT TEMPERATURE OF 30°C (86°F)				IN RACEWAY OR CABLE, BASED ON AMBIENT TEMPERATURE OF 40°C (104°F)		
SIZE	60°C (140°F)	75°C (167°F)	90°C (194°F)	150°C (302°F)	200°C (392°F)	250°C (482°F)
AWG kcmil	TW* UF*	FEPW* RH* RHW* ZW* THW* THWN* XHHW* THHW* USE*	TA,TBS,XHHW-2 SA,RHH*,THW-2 SIS,FEP*,RHW-2 FEPB*,MI,USE-2 THHW*,THWN-2 THHN*,XHH,ZW-2 XHHW*	Z	FEP FEPB PFA	PFAH TFE / NICKEL OR NICKEL-COATED COPPER
14	20*	20*	25*	34	36	39
12	25*	25*	30*	43	45	54
10	30	35*	40*	55	60	73
8	40	50	55	76	83	93
6	55	65	75	96	110	117
4	70	85	95	120	125	148
3	85	100	110	143	152	166
2	95	115	130	160	171	191
1	110	130	150	186	197	215
0	125	150	170	215	229	244
00	145	175	195	251	260	273
000	165	200	225	288	297	308
0000	195	230	260	332	346	361
250	215	255	290	-	-	-
300	240	285	320	-	-	-
350	260	310	350	-	-	-
400	280	335	380	-	-	-
500	320	380	430	-	-	-
600	355	420	475	-	-	-
700	385	460	520	-	-	-
750	400	475	535	-	-	-
800	410	490	555	-	-	-
900	435	520	585	-	-	-
1000	455	545	615	-	-	-
1250	495	590	665	-	-	-
1500	520	625	705	-	-	-
1750	545	650	735	-	-	-
2000	560	665	750	-	-	-

VERCURRENT PROTECTION FOR CONDUCTOR TYPES MARKED (*) WILL NOT XCEED 15 AMPERES FOR SIZE 14 AWG, 20 AMPERES FOR SIZE 12 AWG, AND 0 AMPERES FOR SIZE 10 AWG. (–) FOR DRY LOCATIONS ONLY. SEE 75°C COLUMN FOR WET LOCATIONS.

OR AMBIENT TEMPERATURES OTHER THAN NOTED ABOVE, SEE N.E.C. TABLE 10-16 and 310-18.

ALUMINUM OR COPPER-CLAD ALUMINUM

AMPACITIES OF SINGLE INSULATED CONDUCTORS RATED 0-2000 VOLTS IN FREE AIR					
BASED ON AMBIENT AIR TEMPERATURES OF 30°C (86°F)				40°C (104°F)	
SIZE	60°C (140°F)	75°C (167°F)	90°C (194°F)	150°C (302°F)	BARE
AWG kcmil	TW* UF*	THWN* THHW* XHHW* RHW* THW* RH*	TA,TBS,XHH RHH*,RHW-2 XHHW*,USE-2 THHN*,SA THHW*,THW-2 XHHW-2,ZW-2 THWN-2,SIS	Z	
14	–	–	–	–	25
12	25*	30*	35*	47	30
10	35*	40*	40*	63	35
8	45	55	60	83	55
6	60	75	80	112	75
4	80	100	110	148	100
3	95	115	130	170	120
2	110	135	150	198	135
1	130	155	175	228	160
0	150	180	205	263	185
00	175	210	235	305	215
000	200	240	275	351	250
0000	235	280	315	411	285
250	265	315	355	–	325
300	290	350	395	–	360
350	330	395	445	–	405
400	355	425	480	–	435
500	405	485	545	–	495
600	455	540	615	–	560
700	500	595	675	–	615
750	515	620	700	–	635
800	535	645	725	–	660
900	580	700	785	–	715
1000	625	750	845	–	770
1250	710	855	960	–	–
1500	795	950	1075	–	980
1750	875	1050	1185	–	–
2000	960	1150	1335	–	1215

OVERCURRENT PROTECTION FOR CONDUCTOR TYPES MARKED (*) WILL NOT EXCEED 15 AMPERES FOR SIZE 12 AWG, and 25 AMPS FOR SIZE 10 AWG. (–) DRY LOCATIONS ONLY. SEE 75°C COLUMN FOR WET LOCATIONS.

FOR AMBIENT TEMPERATURES OTHER THAN NOTED ABOVE, SEE N.E.C. TABLE 310-17 and 310-19.

ALUMINUM OR COPPER-CLAD ALUMINUM

AMPACITIES OF NOT MORE THAN THREE SINGLE INSULATED CONDUCTORS RATED 0-2-000 VOLTS				
IN RACEWAY, CABLE, OR EARTH				IN RACEWAY OR CABLE
BASED ON AMBIENT TEMPERATURES OF: 30°C (86°F)				40°C (104°F)
SIZE	60°C (140°F)	75°C (167°F)	90°C (194°F)	150°C (302°F)
AWG kcmil	TW* UF*	THHW* THWN* XHHW* USE* THW* RH*RHW*	TA,TBS,ZW-2 SA,SIS,USE-2 THHW*,XHH XHHW,RHW-2 THHN*THW-2 RHH*THWN-2 XHHW-2	Z
12	20*	20*	25*	30
10	25	30*	35*	44
8	30	40	45	57
6	40	50	60	75
4	55	65	75	94
3	65	75	85	109
2	75	90	100	124
1	85	100	115	145
0	100	120	135	169
00	115	135	150	198
000	130	155	175	227
0000	150	180	205	260
250	170	205	230	-
300	190	230	255	-
350	210	250	280	-
400	225	270	305	-
500	260	310	350	-
600	285	340	385	-
700	310	375	420	-
750	320	385	435	-
800	330	395	450	-
900	355	425	480	-
1000	375	445	500	-
1250	405	485	545	-
1500	435	520	585	-
1750	455	545	615	-
2000	470	560	630	-

OVERCURRENT PROTECTION FOR CONDUCTOR TYPES MARKED (*) WILL NOT EXCEED 15 AMPERES FOR SIZE 12 AWG, AND 25 AMPERES FOR SIZE 10 AWG. (-) FOR WET LOCATIONS ONLY, SEE 75°C COLUMN FOR WET LOCATIONS.

FOR AMBIENT TEMPERATURES OTHER THAN NOTED ABOVE, SEE N.E.C. TABLES 310-16 and 310-18

INSULATION CHART

TRADE NAME	LETTER	MAX. TEMP.	APPLICATION PROVISIONS
FLUORINATED ETHYLENE PROPYLENE	FEP OR FEPB	90°C 194°F 200°C 392°F	DRY AND DAMP LOCATIONS DRY LOCATIONS—SPECIAL APPLICATIONS*
MINERAL INSULATION (METAL SHEATHED)	MI	90°C 194°F 250°C 482°F	DRY AND WET LOCATIONS FOR SPECIAL APPLICATION*
MOISTURE, HEAT, AND OIL RESISTANT THERMOPLASTIC	MTW **	60°C 140°F 90°C 194°F	MACHINE TOOL WIRING IN WET LOCATIONS. NFPA#79-ARTICLE 670 MACHINE TOOL WIRING IN DRY LOCATIONS. NFPA#79-ARTICLE 670
PAPER		85°C 185°F	FOR UNDERGROUND SERVICE CONDUCTORS, OR BY SPECIAL PERMISSION.
PERFLUOROALKOXY	PFA	90°C 194°F 200°C 392°F	DRY AND DAMP LOCATIONS DRY LOCATIONS—SPECIAL APPLICATIONS*
PERFLUOROALKOXY	PFAH	250°C 482°F	DRY LOCATIONS ONLY. ONLY FOR LEADS WITHIN APPARATUS OR WITHIN RACEWAYS CONNECTED TO APPARATUS. (NICKEL OR NICKEL COATED COPPER ONLY.
HEAT-RESISTANT OR CROSS-LINKED SYNTHETIC POLY-MER	RH RHH**	75°C 167°F 90°C 194°F	DRY AND DAMP LOCATIONS DRY AND DAMP LOCATIONS

REFER TO PAGE 103 FOR SPECIAL PROVISIONS AND/OR APPLICATIONS.

INSULATION CHART

TRADE NAME	LETTER	MAX. TEMP.	APPLICATION PROVISIONS
MOISTURE AND HEAT RESISTANT OR CROSS-LINKED SYNTHETIC POLYMER	RHW ** *** RHW-2	75°C 167°F 90°C 194°F	DRY AND WET LOCATIONS. OVER 2,000 VOLT INSULATION SHALL BE OZONE-RESISTANT. DRY AND WET LOCATIONS
SILICONE-ASBESTOS	SA	90°C 194°F 125°C 257°F	DRY AND DAMP LOCATIONS FOR SPECIAL APPLICATIONS*
SYNTHETIC HEAT-RESISTANT	SIS **	90°C 194°F	SWITCHBOARD WIRING ONLY
THERMOPLASTIC AND ASBESTOS	TA	90°C 194°F	SWITCHBOARD WIRING ONLY
THERMOPLASTIC AND FIBROUS OUTER BRAID	TBS	90°C 194°F	SWITCHBOARD WIRING ONLY
EXTENDED POLYTETRAFLUORO-ETHYLENE	TFE	250°C 482°F	DRY LOCATIONS ONLY. ONLY FOR LEADS WITHIN APPARATUS OR WITHIN RACEWAYS CONNECTED TO APPARATUS, (OR AS OPEN WIRING) (NICKLE OR NICKEL COATED COPPER ONLY)
HEAT-RESISTANT THERMOPLASTIC	THHN **	90°C 194°F	DRY AND DAMP LOCATIONS
MOISTURE AND HEAT-RESISTANT THERMOPLASTIC	THHW	75°C 167°F 90°C 194°F	WET LOCATION DRY LOCATION

REFER TO PAGE 103 FOR SPECIAL PROVISIONS AND/OR APPLICATIONS.

INSULATION CHART

TRADE NAME	LETTER	MAX. TEMP.	APPLICATION PROVISIONS
MOISTURE AND HEAT-RESISTANT THERMOPLASTIC	THW *** **	75°C 167°F 90°C 194°F	DRY AND WET LOCATIONS. SPECIAL APPL. WITHIN ELECTRIC DISCHARGE LIGHTING EQUIPMENT. LIMITED TO 1000 OPEN-CIRCUIT VOLTS OR LESS. (SIZE 14-8 ONLY AS PERMITTED IN SECTION 410-31.)
MOISTURE AND HEAT-RESISTANT THERMOPLASTIC	THWN *** **	75°C 167°F	DRY AND WET LOCATIONS
MOISTURE RESISTANT THERMOPLASTIC	TW **	60°C 140°F	DRY AND WET LOCATIONS
UNDERGROUND FEEDER AND BRANCH-CIRCUIT CABLE-SINGLE CONDUCTOR. (FOR TYPE "UF" CABLE EMPLOYING MORE THAN 1 CONDUCTOR. SEE N.E.C. ART. 339)	UF	60°C 140°F 75°C *167°F	SEE ARTICLE 339 N.E.C.
UNDERGROUND SERVICE ENTRANCE CABLE SINGLE CONDUCTOR. (FOR TYPE "USE" CABLE EMPLOYING MORE THAN ONE CONDUCTOR SEE N.E.C. ART. 338.)	USE ***	75°C 167°F	SEE ARTICLE 338 N.E.C.
HEAT-RESISTANT CROSS-LINKED SYNTHETIC POLYMER	XHH**	90°C 194°F	DRY AND DAMP LOCATIONS

REFER TO PAGE 103 FOR SPECIAL PROVISIONS AND/OR APPLICATIONS.

INSULATION CHART

TRADE NAME	LETTER	MAX. TEMP.	APPLICATION PROVISIONS
MOISTURE AND HEAT-RESISTANT CROSS—LINKED SYNTHETIC POLYMER	XHHW *** **	90°C 194°F 75°C 167°F	DRY AND DAMP LOCATIONS WET LOCATIONS
MOISTURE AND HEAT–RESISTANT CROSS-LINKED SYNTHETIC POLYMER	XHHW-2	90°C 194°F	DRY AND WET LOCATIONS
MODIFIED ETHYLENE TETRAFLUORO-ETHYLENE	Z	90°C 194°F 150°C 302°F	DRY AND DAMP LOCATIONS DRY LOCATIONS SPECIAL APPLICATIONS*
MODIFIED ETHYLENE TETRAFLUORO-ETHYLENE	ZW ***	75°C 167°F 90°C 194°F 150°C 302°F	WET LOCATIONS DRY AND DAMP LOCATIONS DRY LOCATIONS-SPECIAL APPLICATIONS*

* WHERE ENVIROMENTAL CONDITIONS REQUIRE MAXIMUM CONDUCTOR OPERATING TEMPERATURES ABOVE 90°C

** INSULATION AND OUTER COVERING THAT MEET THE REQUIREMENTS OF FLAME-RETARDANT, LIMITED SMOKE AND ARE SO LISTED SHALL BE PERMITTED TO BE DESIGNATED LIMITED WITH THE SUFFIX /LS AFTER THE CODE TYPE DESIGNATION.

*** LISTED WIRE TYPES DESIGNATED WITH SUFFIX -2 SUCH AS RHW-2 SHALL BE PERMITTED TO BE USED AT A CONTINUOUS 90°C OPERATING TEMPERATURE WET OR DRY. AMPACITIES OF THESE WIRE TYPES ARE GIVEN IN THE 90°C IN THE APPROPRIATE AMPACITY TABLE.

MAXIMUM NUMBER OF CONDUCTORS IN TRADE SIZES OF CONDUIT OR TUBING

CONDUCTOR		CONDUIT OR TUBING TRADE SIZE (Inches)											
TYPE	SIZE	½	¾	1	1-¼	1-½	2	2-½	3	3-½	4	5	6
RHW,	14	3	6	10	18	25	41	58	90	121	155	–	–
	12	3	5	9	15	21	35	50	77	103	132	–	–
RHH	10	2	4	7	13	18	29	41	64	86	110	–	–
	8	1	2	4	7	9	16	22	35	47	60	94	137
(With Outer Covering)	6	1	1	2	5	6	11	15	24	32	41	64	93
	4	1	1	1	3	5	8	12	18	24	31	50	72
	3	1	1	1	3	4	7	10	16	22	28	44	63
	2	–	1	1	3	4	6	9	14	19	24	38	56
	1	–	1	1	1	3	5	7	11	14	18	29	42
	1/0	–	1	1	1	2	4	6	9	12	16	25	37
	2/0	–	–	1	1	1	3	5	8	11	14	22	32
	3/0	–	–	1	1	1	3	4	7	9	12	19	28
	4/0	–	–	1	1	1	2	4	6	8	10	16	24
	250	–	–	–	1	1	1	3	5	6	8	13	19
	300	–	–	–	1	1	1	3	4	5	7	11	17
	350	–	–	–	1	1	1	2	4	5	6	10	15
	400	–	–	–	1	1	1	1	3	4	6	9	14
	500	–	–	–	1	1	1	1	3	4	5	8	11
	600	–	–	–	–	1	1	1	2	3	4	6	9
	700	–	–	–	–	1	1	1	1	3	3	6	8
	750	–	–	–	–	–	1	1	1	3	3	5	8

MAXIMUM NUMBER OF CONDUCTORS IN TRADE SIZES OF CONDUIT OR TUBING

CONDUCTOR		CONDUIT OR TUBING TRADE SIZE (Inches)											
TYPE	SIZE	½	¾	1	1-¼	1-½	2	2-½	3	3-½	4	5	6
THWN,	14	13	24	39	69	94	154	–	–	–	–	–	–
	12	10	18	29	51	70	114	164	–	–	–	–	–
	10	6	11	18	32	44	73	104	160	–	–	–	–
	8	3	5	9	16	22	36	51	79	106	136	–	–
THHN,	6	1	4	6	11	15	26	37	57	76	98	154	–
FEP (14 Thru 2),	4	1	2	4	7	9	16	22	35	47	60	94	137
FEPB (14 thru 8),	3	1	1	3	6	8	13	19	29	39	51	80	116
PFA (14 thru 4/0),	2	1	1	3	5	7	11	16	25	33	43	67	97
PFAH (14 thru 4/0),	1	–	1	1	3	5	8	12	18	25	32	50	72
Z (14 thru 4/0),	1/0	–	1	1	3	4	7	10	15	21	27	42	61
XHHW (4 thru	2/0	–	1	1	2	3	6	8	13	17	22	35	51
500 kcmil)	3/0	–	1	1	1	3	5	7	11	14	18	29	42
	4/0	–	1	1	1	2	4	6	9	12	15	24	35
	250	–	–	1	1	1	3	4	7	10	12	20	28
	300	–	–	1	1	1	3	4	6	8	11	17	24
	350	–	–	1	1	1	2	3	5	7	9	15	21
	400	–	–	–	1	1	1	3	5	6	8	13	19
	500	–	–	–	1	1	1	2	4	5	7	11	16
	600	–	–	–	1	1	1	1	3	4	5	9	13
	700	–	–	–	–	1	1	1	3	4	5	8	11
	750	–	–	–	–	1	1	1	2	3	4	7	11
XHHW	6	1	3	5	9	13	21	30	47	63	81	128	185
	600	–	–	–	1	1	1	1	3	4	5	9	13
	700	–	–	–	–	1	1	1	3	4	5	7	11
	750	–	–	–	–	1	1	1	2	3	4	7	10

Note 1. This table is for concentric stranded conductors only. For cables with compact conductors, the dimensions in NEC Table 5A shall be used.

Note 2. Conduit fill for conductors with a -2 suffix is the same as for those types without the suffix.

MAXIMUM NUMBER OF CONDUCTORS IN TRADE SIZES OF CONDUIT OR TUBING

CONDUCTOR		CONDUCTOR SIZE IN INCHES											
TYPE	SIZE	½	¾	1	1-¼	1-½	2	2-½	3	3-½	4	5	6
TW, XHHW	14	9	15	25	44	60	99	142	–	–	–	–	–
(14 thru 18)	12	7	12	19	35	47	78	111	171	–	–	–	–
RH (14 + 12)	10	5	9	15	26	36	60	85	131	176	–	–	–
	8	2	4	7	12	17	28	40	62	84	108	–	–
RHW and RHH	14	6	10	16	29	40	65	93	143	192	–	–	–
(WITHOUT OUTER	12	4	8	13	24	32	53	76	117	157	–	–	–
COVERING)	10	4	6	11	19	26	43	61	95	127	163	–	–
RH (10 + 8) THW, THHW	8	1	3	5	10	13	22	32	49	66	85	133	–
TW	6	1	2	4	7	10	16	23	36	48	62	97	141
	4	1	1	3	5	7	12	17	27	36	47	73	106
THW	3	1	1	2	4	6	10	15	23	31	40	63	91
	2	1	1	2	4	5	9	13	20	27	34	54	78
	1	–	1	1	3	4	6	9	14	19	25	39	57
	1/0	–	1	1	2	3	5	8	12	16	21	33	49
	2/0	–	1	1	1	3	5	7	10	14	18	29	41
FEPB (6 thru 2)	3/0	–	1	1	1	2	4	6	9	12	15	24	35
RHW and RHH	4/0	–	–	1	1	1	3	5	7	10	13	20	29
(WITHOUT OUTER COVERING)	250	–	–	1	1	1	2	4	6	8	10	16	23
	300	–	–	1	1	1	2	3	5	7	9	14	20
	350	–	–	–	1	1	1	3	4	6	8	12	18
	400	–	–	–	1	1	1	2	4	5	7	11	16
RH, THHW	500	–	–	–	1	1	1	1	3	4	6	9	14
	600	–	–	–	–	1	1	1	3	4	5	7	11
	700	–	–	–	–	1	1	1	2	3	4	7	10
	750	–	–	–	–	1	1	1	2	3	4	6	9

NOTE: THIS TABLE IS FOR THE CONCENTRIC STRANDED CONDUCTORS ONLY. FOR CABLES WITH THE COMPACT CONDUCTORS, THE DIMENSIONS IN TABLE 5A SHALL BE USED.

MAXIMUM NUMBER OF FIXTURE WIRES IN TRADE SIZES OF CONDUIT OR TUBING (40 PERCENT FILL BASED ON INDIVIDUAL DIAMETERS)

CONDUIT TRADE SIZE	1/2"			3/4"			1"			1-1/4"			1-1/2"			2"		
WIRE TYPES	18	16	14	18	16	14	18	16	14	18	16	14	18	16	14	18	16	14
PTF, PTFF, PGFF, PGF, PFF, PF, PAF, PAFF, ZF, ZFF	23	18	14	40	31	24	65	50	39	115	90	70	157	122	95	257	200	156
TFFN, TFN	19	15		34	26		55	43		97	76		132	104		216	169	
SF-1	16			29			47			83			114			186		
SFF-1	15			26			43			76			104			169		
TF	11	10		20	18		32	30		57	53		79	72		129	118	
RFH-1	11			20			32			57			79			129		
TFF	11	10		20	17		32	27		56	49		77	66		126	109	
SFF-2	9	7	6	16	12	10	27	20	17	47	36	30	65	49	42	106	81	68
SF-2	9	8	6	16	14	11	27	23	18	47	40	32	65	55	43	106	90	71
FFH-2	9	7		15	12		25	19		44	34		60	46		99	75	
RFH-2	7	5		12	10		20	16		36	28		49	38		80	62	
*AF	11	9	7	19	16	12	31	26	20	55	46	36	75	63	49	123	104	81
**KF-1, KFF-1, KF-2, KFF-2	36	32	22	64	55	39	103	89	63	182	158	111	248	216	152	406	353	248
WIRE TYPES	12		10	12		10	12		10	12		10	12		10	12		10
*AF	4		3	7		5	11		8	19		15	27		20	44		34
**KF-1, KFF-1, KF-2, KFF-2	14		9	25		17	41		28	73		49	100		67	163		110

METAL BOXES

BOX DIMENSION, INCHES TRADE SIZE OR TYPE	MIN. CU. IN. CAPACITY	MAXIMUM NUMBER OF CONDUCTORS						
		No.18	No.16	No.14	No.12	No.10	No.8	No.6
4 x 1-1/4 ROUND OR OCTAGONAL	12.5	8	7	6	5	5	4	2
4 x 1-1/2 ROUND OR OCTAGONAL	15.5	10	8	7	6	6	5	3
4 x 2-1/8 ROUND OR OCTAGONAL	21.5	14	12	10	9	8	7	4
4 x 1-1/4 SQUARE	18.0	12	10	9	8	7	6	3
4 x 1-1/2 SQUARE	21.0	14	12	10	9	8	7	4
4 x 2-1/8 SQUARE	30.3	20	17	15	13	12	10	6
4-11/16 x 1-1/4 SQUARE	25.5	17	14	12	11	10	8	5
4-11/16 x 1-1/2 SQUARE	29.5	19	16	14	13	11	9	5
4-11/16 x 2-1/8 SQUARE	42.0	28	24	21	18	16	14	8
3 x 2 x 1-1/2 DEVICE	7.5	5	4	3	3	3	2	1
3 x 2 x 2 DEVICE	10.0	6	5	5	4	4	3	2
3 x 2 x 2-1/4 DEVICE	10.5	7	6	5	4	4	3	2
3 x 2 x 2-1/2 DEVICE	12.5	8	7	6	5	5	4	2
3 x 2 x 2-3/4 DEVICE	14.0	9	8	7	6	5	4	2
3 x 2 x 3-1/2 DEVICE	18.0	12	10	9	8	7	6	3
4 x 2-1/8 x 1-1/2 DEVICE	10.3	6	5	5	4	4	3	2
4 x 2-1/8 x 1-7/8 DEVICE	13.0	8	7	6	5	5	4	2
4 x 2-1/8 x 2-1/8 DEVICE	14.5	9	8	7	6	5	4	2
3-3/4 x 2 x 2-1/2 MASONRY BOX/GANG	14.0	9	8	7	6	5	4	2
3-3/4 x 2 x 3-1/2 MASONRY BOX/GANG	21.0	14	12	10	9	8	7	4
FS-MINIMUM INTERNAL DEPTH 1-3/4 SINGLE COVER/GANG	13.5	9	7	6	6	5	4	2
FD-MINIMUM INTERNAL DEPTH 2-3/8 SINGLE COVER/GANG	18.0	12	10	9	8	7	6	3
FS-MINIMUM INTERNAL DEPTH 1-3/4 MULTIPLE COVER/GANG	18.0	12	10	9	8	7	6	3
FD-MINIMUM INTERNAL DEPTH 2-3/8 MULTIPLE COVER/GANG	24.0	16	13	12	10	9	8	4

MINIMUM COVER REQUIREMENTS 0-600 VOLTS, NOMINAL

COVER IS DEFINED AS THE DISTANCE BETWEEN THE TOP SURFACE OF DIRECT BURIAL CABLE, CONDUIT, OR OTHER RACEWAYS AND THE FINISHED SURFACE.

WIRING METHOD	MINIMUM BURIAL (INCHES)
DIRECT BURIAL CABLES	24
RIGID METAL CONDUIT	6
INTERMEDIATE METAL CONDUIT	6
RIGID NONMETALLIC CONDUIT (APPROVED FOR DIRECT BURIAL WITHOUT CONCRETE ENCASEMENT)	18

FOR MOST LOCATIONS. FOR COMPLETE DETAILS REFER TO NATIONAL ELECTRICAL CODE ® TABLE 300-5 FOR EXCEPTIONS SUCH AS HIGHWAYS, AIRPORTS, DRIVEWAYS, PARKING LOTS, ECT.

VOLUME REQUIRED PER CONDUCTOR

SIZE OF CONDUCTOR	FREE SPACE WITHIN BOX FOR EACH CONDUCTOR
No. 18	1.5 CUBIC INCHES
No. 16	1.75 CUBIC INCHES
No. 14	2 CUBIC INCHES
No. 12	2.25 CUBIC INCHES
No. 10	2.5 CUBIC INCHES
No. 8	3 CUBIC INCHES
No. 6	5 CUBIC INCHES

FOR COMPLETE DETAILS SEE NEC 370-16.

VERTICAL CONDUCTOR SUPPORTS

AWG or Circular-Mil Size of Wire	Support of Conductors in Vertical Raceways	CONDUCTORS Aluminum or Copper-Clad Aluminum	 Copper
18 AWG through 8 AWG	Not greater than	100 feet	100feet
6 AWG through 1/0 AWG	Not greater than	200 feet	100feet
2/0 AWG through 4/0 AWG	Not greater than	180 feet	80feet
Over 4/0 AWG through 350 kcmil	Not greater than	135 feet	60feet
Over 350 kcmil through 500 kcmil	Not greater than	120 feet	50feet
Over 500 kcmil through 750 kcmil	Not greater than	95 feet	40feet
Over 750 kcmil	Not greater than	85 feet	35feet

For SI units: one foot = 0.3048 meter.

MINIMUM DEPTH OF CLEAR WORKING SPACE IN FRONT OF ELECTRICAL EQUIPMENT

NOMINAL VOLTAGE TO GROUND	CONDITIONS		
	1	2	3
	Minimum clear distance (feet)		
0-150	3	3	3
151-600	3	3-1/2	4
601-2500	3	4	5
2501-9000	4	5	6
9001-25,000	5	6	9
25,001-75 kV	6	8	10
Above 75 kV	8	10	12

Where the "Conditions" are as follows:

1. Exposed live parts on one side and no live or grounded parts on the other side of the working space, or exposed live parts on both sides effectively guarded by suitable wood or other insulating materials. Insulated wire or insulated busbars operating at not over 300 volts shall not be considered live parts.

2. Exposed live parts on one side and grounded parts on the other side.

3. Exposed live parts on both sides of the work space (not guarded as provided in Condition 1) with the operator between.

★ See NEC 110 and 110-A for exceptions.

MINIMUM CLEARANCE OF LIVE PARTS

NOMINAL VOLTAGE RATING, KV	IMPULSE WITHSTAND, B.I.L. KV		*MINIMUM CLEARANCE OF LIVE PARTS, INCHES			
			PHASE-TO-PHASE		PHASE-TO-GROUND	
	INDOORS	OUTDOORS	INDOORS	OUTDOORS	INDOORS	OUTDOORS
2.4-4.16	60	95	4.5	7	3.0	6
7.2	75	95	5.5	7	4.0	6
13.8	95	110	7.5	12	5.0	7
14.4	110	110	9.0	12	6.5	7
23	125	150	10.5	15	7.5	10
34.5	150	150	12.5	15	9.5	10
	200	200	18.0	18	13.0	13
46		200		18		13
		250		21		17
69		250		21		17
		350		31		25
115		550		53		42
138		550		53		42
		650		63		50
161		650		63		50
		750		72		58
230		750		72		58
		900		89		71
		1050		105		83

OR SI UNITS: ONE INCH = 25.4 MILLIMETERS.
THE VALUES GIVEN ARE THE MINIMUM CLEARANCE FOR RIGID PARTS AND BARE CONDUCTORS UNDER FAVORABLE SERVICE CONDITIONS. THEY SHALL BE INCREASED FOR CONDUCTOR MOVEMENT OR UNDER UNFAVORABLE SERVICE CONDITIONS, OR WHEREVER SPACE LIMITATIONS PERMIT. THE SELECTION OF THE ASSOCIATED IMPULSE WITHSTAND VOLTAGE FOR A PARTICULAR SYSTEM VOLTAGE IS DETERMINED BY THE CHARACTERISTICS OF THE SURGE PROTECTIVE EQUIPMENT.

MINIMUM SIZE EQUIPMENT GROUNDING CONDUCTORS FOR GROUNDING RACEWAY AND EQUIPMENT

RATING OR SETTING OF AUTOMATIC OVERCURRENT DEVICE IN CIRCUIT AHEAD OF EQUIPMENT, CONDUIT, ETC., NOT EXCEEDING (AMPERES)	SIZE	
	COPPER WIRE NO.	ALUMINUM OR COPPER-CLAD ALUMINUM WIRE NO.
15	14	12
20	12	10
30	10	8
40	10	8
60	10	8
100	8	6
200	6	4
300	4	2
400	3	1
500	2	1/0
600	1	2/0
800	1/0	3/0
1000	2/0	4/0
1200	3/0	250 kcmil
1600	4/0	350 kcmil
2000	250 kcmil	400 kcmil
2500	350 kcmil	600 kcmil
3000	400 kcmil	600 kcmil
4000	500 kcmil	800 kcmil
5000	700 kcmil	1200 kcmil
6000	800 kcmil	1200 kcmil

GROUNDING ELECTRODE CONDUCTOR FOR AC SYSTEMS

SIZE OF LARGEST SERVICE-ENTRANCE CONDUCTOR OR EQUIVALENT AREA FOR PARALLEL CONDUCTORS		SIZE OF GROUNDING ELECTRODE CONDUCTOR	
COPPER	ALUMINUM OR COPPER-CLAD ALUMINUM	COPPER	*ALUMINUM OR COPPER-CLAD ALUMINUM
2 OR SMALLER	1/0 OR SMALLER	8	6
1 OR 1/0	2/0 OR 3/0	6	4
2/0 OR 3/0	4/0 OR 250 kcmil	4	2
OVER 3/0 THRU 350 kcmil	OVER 250 THRU THRU 500 kcmil	2	1/0
OVER 350 kcmil THRU 600 kcmil	OVER 500 kcmil THRU 900 kcmil	1/0	3/0
OVER 600 kcmil THRU 1100 kcmil	OVER 900 kcmil THRU 1750 kcmil	2/0	4/0
OVER 1100 kcmil	OVER 1750 kcmil	3/0	250 kcmil

*WHERE THERE ARE NO SERVICE-ENTRANCE CONDUCTORS, THE GROUNDING ELECTRODE CONDUCTOR SIZE SHALL BE DETERMINED BY THE EQUIVALENT SIZE OF THE LARGEST SERVICE-ENTRANCE CONDUCTOR REQUIRED FOR THE LOAD TO BE SERVED.

ELECTRICAL SYMBOLS

WALL CEILING

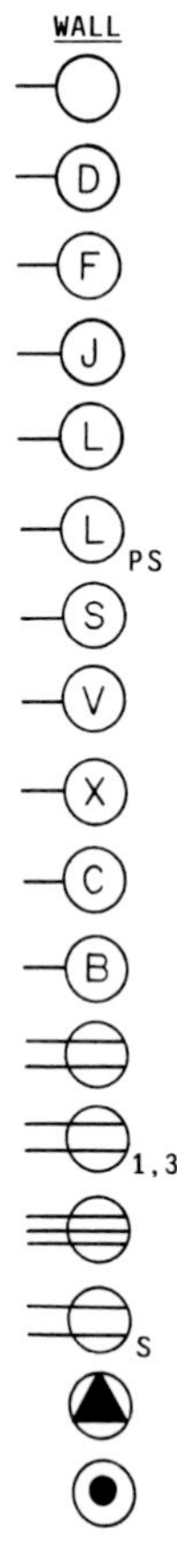

- OUTLET
- DROP CORD
- FAN OUTLET
- JUNCTION BOX
- LAMP HOLDER
- LAMP HOLDER WITH PULL SWITCH
- PULL SWITCH
- VAPOR DISCHARGE SWITCH
- EXIT LIGHT
- CLOCK OUTLET
- BLANKED OUTLET
- DUPLEX CONVENIENCE OUTLET
- SINGLE, TRIPLEX, ETC.
- RANGE OUTLET
- SWITCH AND CONVENIENCE OUTLET
- SPECIAL PURPOSE OUTLET
- FLOOR OUTLET

SWITCH OUTLETS

- S SINGLE POLE SWITCH
- S_2 DOUBLE POLE SWITCH
- S_3 THREE WAY SWITCH
- S_4 FOUR WAY SWITCH
- S_D AUTOMATIC DOOR SWITCH
- S_E ELECTROLIER SWITCH
- S_P SWITCH AND PILOT LAMP
- S_K KEY OPERATED SWITCH
- S_{CB} CIRCUIT BREAKER
- S_{WCB} WEATHER PROOF CIRCUIT BREAKER
- S_{MC} MOMENTARY CONTACT SWITCH
- S_{RC} REMOTE CONTROL SWITCH
- S_{WP} WEATHER PROOF SWITCH
- S_F FUSED SWITCH
- S_{WPF} WEATHER PROOF FUSED SWITCH
- LIGHTING SWITCH
- POWER PANEL

Reproduced from American Standard IEEE.

ELECTRICAL SYMBOLS

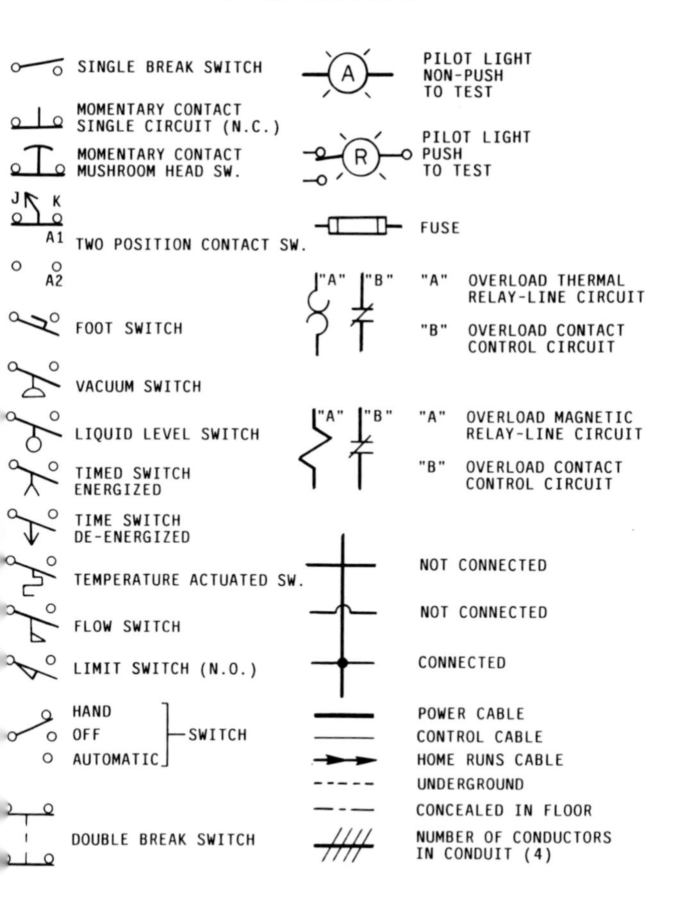

eproduced from American Standard IEEE.

ELECTRICAL SYMBOLS

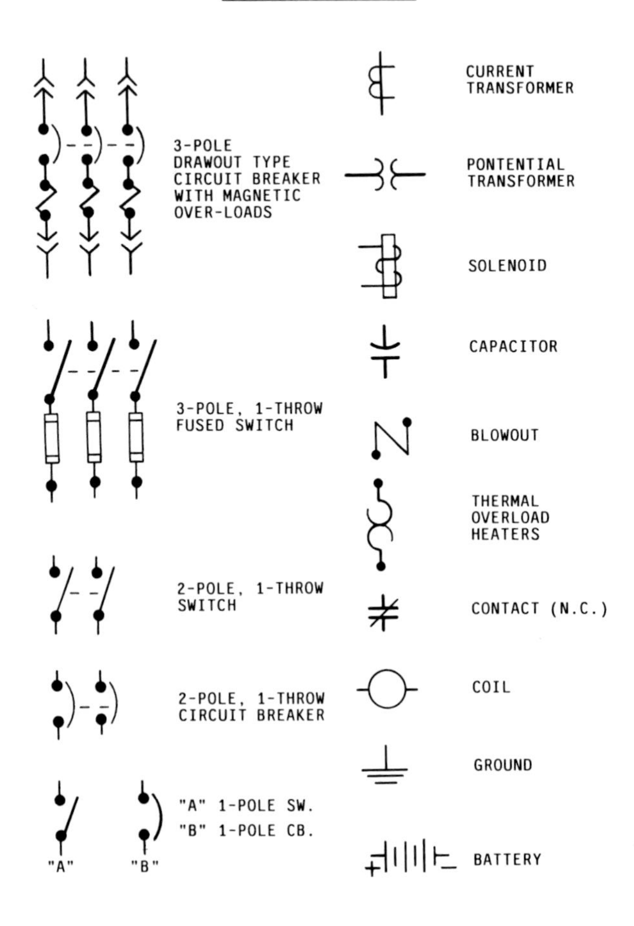

Reproduced from American Standard IEEE.

Wiring Diagrams For NEMA Configurations

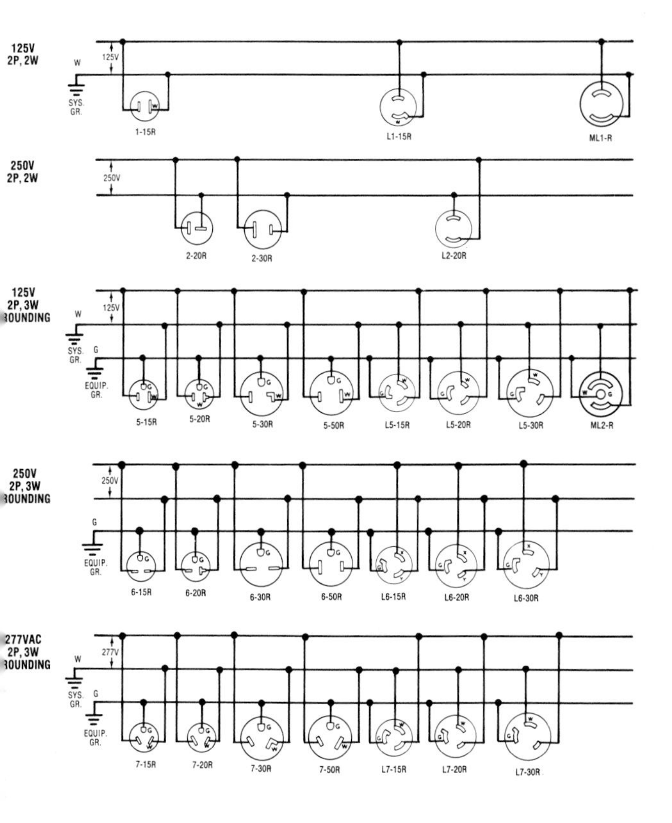

Courtesy of Cooper Industries, Inc. - Arrow Hart Wiring Devices

Wiring Diagrams
For NEMA Configurations

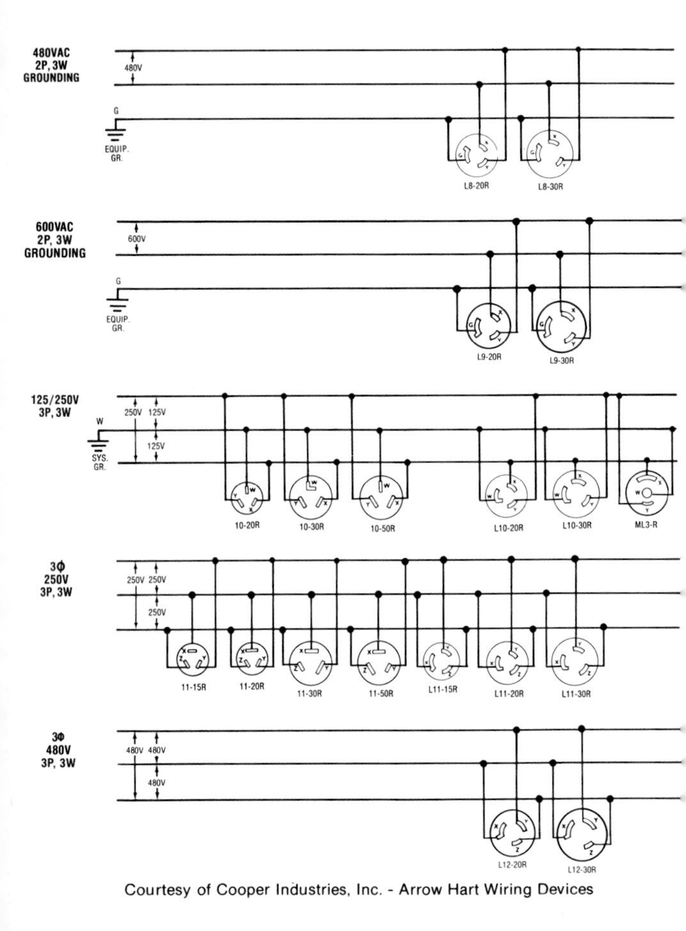

Courtesy of Cooper Industries, Inc. - Arrow Hart Wiring Devices

Wiring Diagrams For NEMA Configurations

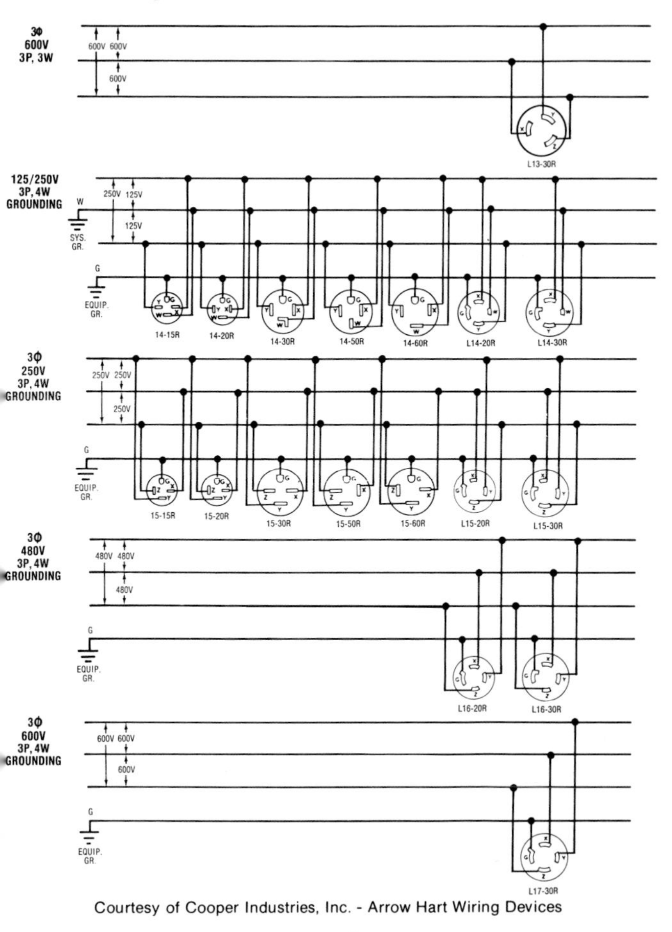

Courtesy of Cooper Industries, Inc. - Arrow Hart Wiring Devices

Wiring Diagrams For NEMA Configurations

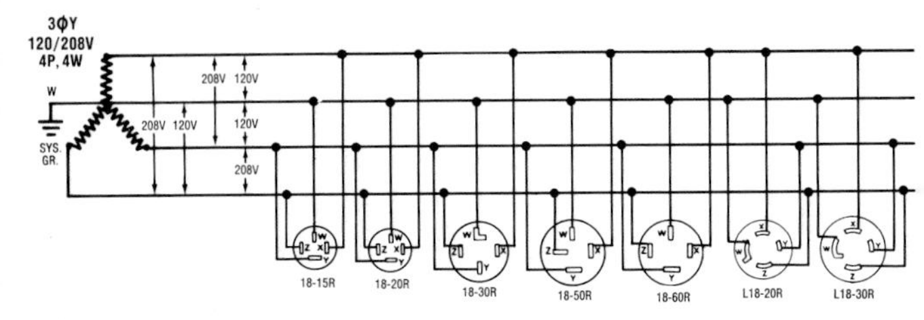

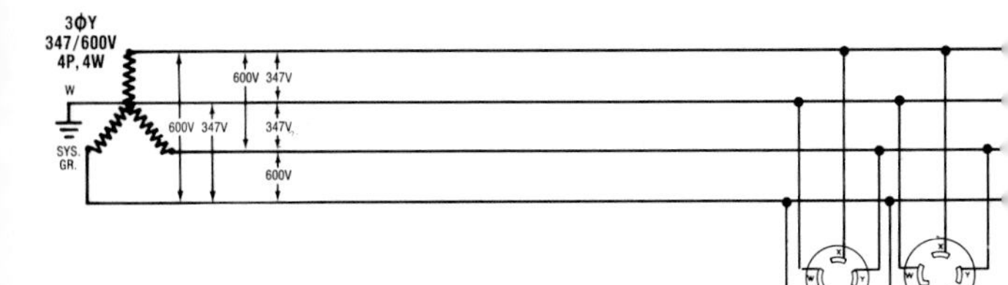

Courtesy of Cooper Industries, Inc. - Arrow Hart Wiring Devices

Wiring Diagrams For NEMA Configurations

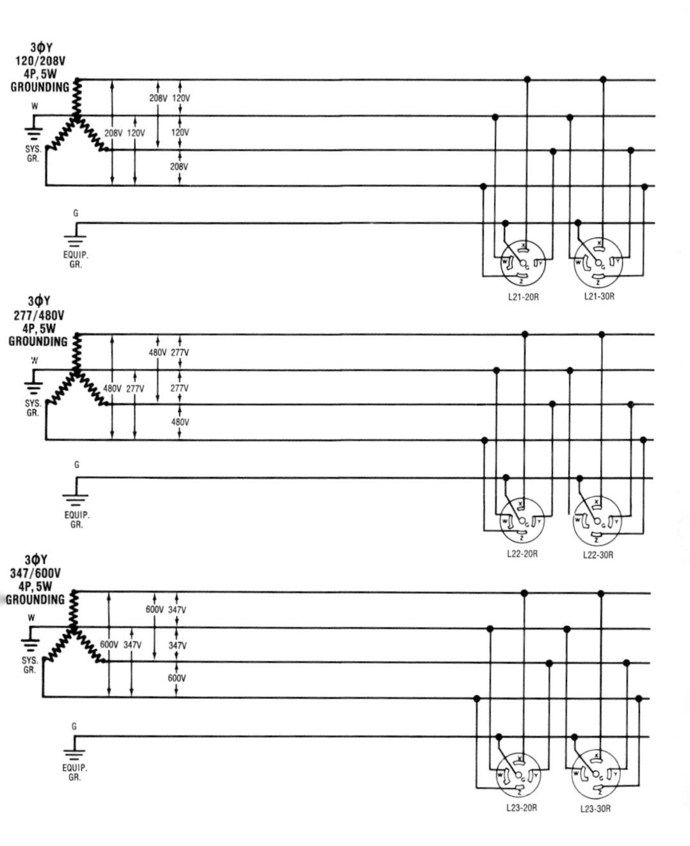

Courtesy of Cooper Industries, Inc. - Arrow Hart Wiring Devices

HAND SIGNALS FOR CRANES AND CHERRY PICKERS

STOP

DOG EVERYTHING

EMERGENCY STOP

TRAVEL

TRAVEL BOTH TRACKS (CRAWLER CRANES ONLY)

TRAVEL ONE TRACK (CRAWLERS)

RETRACT BOOM

EXTEND BOOM

SWING BOOM

HAND SIGNALS FOR CRANES AND CHERRY PICKERS

RAISE LOAD

LOWER LOAD

MAIN HOIST

MOVE SLOWLY

RAISE BOOM AND LOWER LOAD (FLEX FINGERS)

LOWER BOOM AND RAISE LOAD (FLEX FINGERS)

USE WHIP LINE

BOOM UP

BOOM DOWN

USEFUL KNOTS

BOWLINE

RUNNING BOWLINE

BOWLINE ON THE BIGHT

CLOVE HITCH

SHEEPSHANK

ROLLING HITCH

SINGLE BLACKWALL HITCH

CATSPAW

DOUBLE BLACKWALL HITCH

SQUARE KNOT

TIMBER HITCH WITH HALF HITCH

SINGLE SHEET BEND

GENERAL DIRECTIONS FOR FIRST AID:

While help is being summoned, do the following:
1) Minimize further injury—move victim only if necessary for safety reasons.
2) Control severe bleeding
3) Maintain an open airway and give **Rescue Breathing** or CPR if necessary.
4) Treat for shock.

URGENT CARE:

BLEEDING

First Aid:
1) Direct Pressure and Elevation:
   * Place dressing and apply direct pressure directly over the wound, then elevate above the level of the heart, unless there is evidence of a fracture.
2) Apply pressure bandage:
   * Wrap bandage snugly over the dressing.
3) Pressure Points
   * If bleeding doesn't stop after direct pressure, elevation, and the pressure bandage, compress the pressure point.
   * Arm: Use the brachial artery—pushing the artery against the upper arm bone.
   * Leg: Apply pressure on femoral artery, pushing it against the pelvic bone.

Nosebleed:
   * To control a nosebleed, have the victim lean forward and pinch the nostrils together until bleeding stops.

POISONING

Signals: Vomiting, heavy labored breathing, sudden onset of pain or illness, burns or odor around the lips or mouth, unusual behavior.

FIRST Aid:
   *** If you think someone has been poisoned, call your poison control center or local emergency number and follow their directions.**

   If **conscious:**
   * Call Poison Control and try to identify the poison - be prepared to inform poison center of the type of poison, when incident occurred, victims age, symptoms, and how much poison may have been ingested, inhaled, absorbed, or injected.

   If **unconscious** or nauseous:
   1) Position victim on side and monitor vital signs (i.e. pulse and breathing).
   2) Call Poison Control and identify poison.
   3) DO NOT give anything by mouth.

SHOCK

Signals: Cool, moist, pale, bluish skin, weak rapid pulse (over 100),nausea, rate of breathing increased, apathetic.

First Aid:
1) Maintain open airway, have victim lie down
2) Maintain normal body temperature(98.6). If too hot, cool down, and if too cold, use blankets,over and under, to warm the victim.

BURNS

Signals: Small, thin (surface) burns or large, thin burns: redness, pain, and swelling. Deep burns: blisters, deep tissue destruction, charred appearance.

First Aid:
1) Stop the burning — put out flames or remove the victim from the source of the burn.
2) Cool the burn —run or pour cool water on burn. Immerse if possible. Cool until pain is reduced.
3) Cover the burn — Use dry, sterile dressing and bandage.
4) Keep victim comfortable as possible from being chilled or over heated.

Chemical burn -- must be flushed with large amounts of water until EMS arrives.
Electrical burn -- make sure power is turned off before touching the victim.

ELECTRICAL SHOCK

Signals: Unconsciousness, absence of breathing & pulse

First Aid:

1) TURN OFF THE POWER SOURCE — Call EMS
   (DO NOT approach victim **until** power has been turned off.)
2) DO NOT move a victim of electrical injury unless there is immediate danger.
3) Administer rescue breathing or CPR if necessary.
4) Treat for Shock.
5) Check for other injuries and monitor victim till medical help arrives.

FROSTBITE

Signals: Flushed, white, or gray skin. Pain. The nose, cheeks, ears, fingers, and toes are most likely to be affected. Pain may be felt early and then subside. Blisters may appear later.

First Aid:

1) Cover the frozen part. Loosen restrictive clothing or boots.
2) Bring victim indoors ASAP.
3) Give the victim a warm drink. (DO NOT give alcoholic beverages, tea, or coffee)
4) Immerse frozen part in warm water (102°-105°), or wrap in a sheet and warm blankets. DO NOT rewarm if there is a possibility of refreezing.
5) Remove from water and discontinue warming once part becomes flushed.
6) After thawing, the victim should try to move the injured area a little.
7) Elevate the injured area and protect from further injury.
8) DO NOT rub the frozen part. DO NOT break the blisters. DO NOT use extreme or dry heat to rewarm the part.
9) If fingers or toes are involved, place dry, sterile gauze between them when bandaging.

HYPOTHERMIA

Signals: Lowered body core temperature. Persistent shivering, lips may be blue, slow slurred speech, memory lapses. Most cases occur when air temperature ranges from 30°-50° or water temperature is below 70°F

First Aid:

1) Move victim to shelter and remove wet clothing if necessary.
2) Rewarm victim with blankets or body-to-body contact in sleeping bag.
3) If victim is conscious and able to swallow, give warm liquids.
4) Keep victim warm & quiet.
5) DO NOT give alcoholic beverages, or beverages containing caffeine.
6) Constantly monitor victim and give Rescue Breathing and CPR if necessary.

HEAT EXHAUSTION/HEAT STROKE

Signals: Heat Exhaustion: Pale clammy skin, profuse perspiration, weakness, nausea, headache. Heat Stroke: Hot dry red skin, no perspiration, rapid & strong pulse. High body temperature (105°+) This is an immediate life threatening emergency.

First Aid:

1) Get the victim out of the heat.
2) Loosen tight clothing or restrictive clothing.
3) Remove perspiration soaked clothing.
4) Apply cool, wet cloths to the skin.
5) Fan the victim.
6) If victim is conscious give cool water to drink.
7) Call for an ambulance if victim refuses water, vomits, or starts to lose consciousness.

RESCUE BREATHING

1. Check the victim.

Tap and shout, " Are you okay?", to see if the person responds.

If no response ...

2. CALL EMS

3. CARE FOR THE VICTIM

Step 1: Look, listen and feel for breathing for about 5 seconds.

If the person is not breathing or you can't tell ..

Step 2: Position victim on back, while supporting head and neck.

Step 3: Tilt head back and lift chin.

Step 4: Look, listen, and feel for breathing for about 5 seconds.

If not breathing ...

Step 5: Give two slow gentle breaths.

Step 6: Check pulse for 5 to 10 seconds.

Step 7: Check for severe bleeding.

4. GIVE RESCUE BREATHING

If pulse is present but person is still not breathing ...

Step 1: Give one slow breath about every 5 seconds. Do this for about 1 minute (12 breaths).

Step 2: Recheck pulse and breathing about every minute.

Continue rescue breathing as long as pulse is present but person is not breathing.

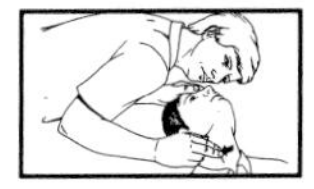

5. BEGIN CARDIOPULMONARY RESUSCITATION (CPR)

If there is no pulse and no breathing.

FIRST AID FOR CHOKING

1. Check the victim.

When an adult is choking:

Step 1: Ask, **" Are you choking?"** If victim cannot cough, speak, or breathe, is coughing weakly or is making high-pitched noises...

Step 2: Shout, **"HELP!"**

Step 3: **Phone EMS** for help —Send someone to call for an ambulance.

Step 4: **Do abdominal thrusts:**

A. Wrap your arms around victim's waist. Make a fist. Place thumbside of fist against middle of abdomen just above the navel. Grasp fist with other hand.

If victim becomes unconscious, lower victim to floor.

B. Give quick, upward thrusts. **Repeat until object is coughed up or person becomes unconscious.**

IF VICTIM BECOMES UNCONSCIOUS, LOWER VICTIM TO FLOOR.

Step 5: Do a Finger Sweep — Lift jaw and tongue, do a finger sweep to remove any obstruction.

Step 6: Open Airway — Tilt head back and lift chin.

Step 7: **Attempt to give breaths.** With head tilted back and chin lifted, pinch nose shut. Give two slow breaths for 1 1/2 - 2 seconds each.

If air won't go in...

Step 8: **Give up to 5 Abdominal thrusts — If air won't go in ...** Place heel of one hand against middle of victim's abdomen just above the navel. Place other hand on top of first hand. Press into abdomen with up to five quick upward thrusts.

Repeat breaths, thrusts, and sweeps until breaths go in or victim starts to breathe.